Elastic Instability Phenomena

Elastic Instability Phenomena

J. M. T. Thompson

Department of Civil Engineering
University College London

and

G. W. Hunt

Department of Civil Engineering
Imperial College London

A Wiley–Interscience Publication

JOHN WILEY AND SONS

Chichester · New York · Brisbane · Toronto · Singapore

69946140

ENGINEERING

British Library Cataloguing in Publication Data:

Thompson, J. M. T.
  Elastic instability phenomena.
  1. Structural dynamics
  I. Title   II. Hunt, G. W.
  624.1'76   TA656

  ISBN 0 471 90279 9

Library of Congress Cataloging in Publication Data:

Thompson, J. M. T.
  Elastic instability phenomena.

  'A Wiley–Interscience publication.'
  Bibliography: p.
  Includes index.
  1. Structural stability.   2. Buckling (Mechanics)
3. Elasticity.   I. Hunt, G. W.   II. Title.
TA656.T48 1984   624.1'71   83–14514
ISBN 0 471 90279 9

Photosetting by Thomson Press (India) Limited, New Delhi
and printed by Page Bros. (Norwich) Limited.

Contents

Preface.   ix

1   The General Conservative System.   1
   1.1   Generalized coordinates   1
   1.2   Lagrange equations   3
   1.3   Statical equilibrium   5
   1.4   Stability definition   7
   1.5   Energy theorems.   9
   1.6   Conditions for a minimum   12
   1.7   Stability of a critical state.   21
   1.8   Linear vibrations   23

2   Vibration and Buckling of Beams and Struts.   27
   2.1   Beam formulation.   27
   2.2   Modal expansions.   29
   2.3   Complete harmonic analysis of a column   31
   2.4   Vibration of a cantilever in two modes   33
   2.5   The finite element method.   34
   2.6   Buckling of a strut in four modes   37

3   Loads and Imperfections.   40
   3.1   Related fields of applicability   40
   3.2   Elimination of passive coordinates   42
   3.3   Loss of stability under load   43
   3.4   Imperfections and perturbed bifurcations.   47
   3.5   Elimination of passive controls   53
   3.6   Catastrophes of Thom and Zeeman   56
   3.7   Bifurcations of Golubitsky and Schaeffer.   58

4   Distinct Buckling Phenomena   60
   4.1   The fold singularity   60

4.2 Snap-buckling at a limit point............................ 63
4.3 Asymmetric point of bifurcation 64
4.4 Routes through the fold 72
4.5 The cusp singularity..................................... 73
4.6 Stable-symmetric point of bifurcation..................... 75
4.7 Unstable-symmetric point of bifurcation 81
4.8 Routes through the cusp.................................. 83
4.9 Higher-order uni-modal singularities 86

5 Buckling and Imperfection-Sensitivity of Arches................... 91
5.1 Simplification via inextensibility 91
5.2 Strain energy with arbitrary pre-stress..................... 91
5.3 Expansion in Fourier harmonics........................... 93
5.4 The constraint condition.................................. 95
5.5 Linear eigenvalue analysis................................ 98
5.6 Post-buckling analysis................................... 101
5.7 The real perfect response................................ 103
5.8 Imperfection-sensitivity analysis.......................... 104
5.9 Comparison with experiments 105
5.10 Tilt as a second imperfection 107
5.11 Convolutions in the symmetric response 108

6 Interactive Buckling Phenomena 112
6.1 The guyed cantilever 112
6.2 Semi-symmetric points of bifurcation 118
6.3 Imperfection-sensitivity surfaces......................... 126
6.4 Fully asymmetric points of bifurcation.................... 133
6.5 Routes through the umbilic catastrophes.................. 135
6.6 Higher-order two-mode singularities...................... 140

7 Comprehensive Bifurcation Analysis 147
7.1 Bifurcational formalism.................................. 148
7.2 Elimination of passive coordinates 150
7.3 Perturbation analysis.................................... 153
7.4 Generalized imperfections............................... 157
7.5 Generalized loads....................................... 158
7.6 Illustration of compound semi-symmetric buckling........... 160

8 Buckling of Plates and Shells................................ 164
8.1 Analysis using principal coordinates...................... 165
8.2 Symmetry and optimization 167
8.3 The Euler strut 169
8.4 Post-buckling of a compressed plate....................... 170
8.5 Interactive buckling of stiffened structures................. 175
8.6 The axially compressed cylindrical shell................... 180

8.7 The externally pressurized spherical shell.................... 184
8.8 Rigid and semi-rigid laboratory loading devices.............. 188

Appendix: Proof of the Lagrange and Hamilton Equations 195

References.. 198

Index 204

Preface

The present work treats the buckling and post-buckling of engineering structures and components, with an emphasis on the important non-linear features of behaviour. The subject is developed from the foundation of structural dynamics, and concentrates on the buckling of structures under conservative loading, including the significant imperfection-sensitivity of stiffened plates and shells.

The book is intended for undergraduate and post-graduate courses in civil, mechanical, marine, and aerospace engineering, and includes material that has been taught for a number of years to undergraduate and MSc students of Civil Engineering at University College and Imperial College, London. It can be seen as a complement to two earlier books by the same authors. The first, *A General Theory of Elastic Stability*,[1] by J. M. T. Thompson and G. W. Hunt, published by John Wiley in 1973, was an advanced text on non-linear bifurcation theory, dealing with the elastic buckling and post-buckling of conservative mechanical systems and structures. The second, *Instabilities and Catastrophes in Science and Engineering*,[2] by J. M. T. Thompson, published by John Wiley in 1982, described in a more general, less technical way a wide variety of instability phenomena drawn from the breadth of science and technology. It dealt with the dynamic flutter instabilities of non-conservative systems in addition to the static buckling instabilities of conservative systems, and related both to recent developments in dynamical systems theory, of which catastrophe theory is a part.

This book returns to the theme and spirit of the first of those works,[1] which it complements and completes in a number of important ways. It is restricted to conservative systems, and presents *first*, the underlying *dynamical framework* in which instability problems should properly be viewed: these foundations, based on the Lagrange equations, were not presented explicitly by Thompson and Hunt.[1] *Second*, it relates the engineering theory of elastic stability to recent advances in *singularity theory* which come under the name of catastrophe theory. This broadens the context of the work, and adds some useful new points of view, including the important mathematical concept of a structurally stable topology. A significant *third* contribution is a full extension of the earlier outline of *interactive buckling* at compound branching points, in which the topological view is found to contribute non-trivially, leading to a deeper understanding than had hitherto been available.

Closely related as it is to our earlier books, we have nevertheless tried to make this book as self-contained as possible. This has inevitably involved a little repetition and overlapping of our previous work, but we have kept this to a minimum by presenting, whenever possible, an alternative view using different illustrative examples.

Chapter 1 lays the conceptual foundations of stability theory. The principles are developed gradually but systematically through physically understandable examples, starting with the simple oscillations of a pendulum. It gives a thorough treatment of the general conservative mechanical system based on the Lagrange equations. Stability is defined in the manner of Liapunov by the form of the phase trajectories close to an equilibrium state, and the powerful energy theorems are established. Linear and non-linear conditions for a minimum of the total potential energy are presented, leading to the definition of stability coefficients and normal modes of vibration.

Chapter 2 presents multi-mode linear eigenvalue analyses of beams and columns. Fourier expansions yield complete closed-form results for the vibration and buckling of struts, while discontinuous Rayleigh–Ritz modal analyses serve as an introduction to finite element methods.

Chapter 3 introduces loads and imperfections as control parameters modifying the potential energy of the system. It looks briefly at related fields that are covered by the present theory,[2] including the thermodynamics of stars and the fracture of crystals. The elimination of passive coordinates is presented, showing how the buckling modes alone govern the incipient instability of a structure. Evolution under an increasing load leads to a bifurcational view of instability phenomena, and four common distinct critical points are delineated, along with two basic theorems recently proved by mathematical topologists. Imperfections perturb these bifurcations, generating imperfection-sensitivity phenomena. Mathematical ideas of topological stability allow the numeration and identification of active control parameters (loads, imperfections, and geometrical parameters) essential for the full description of a given singularity. The chapter ends by looking at the *catastrophe theory* classifications of Thom and Zeeman[3–5] and the related but finer *bifurcational classifications* of Golubitsky and Schaeffer.[6]

Chapter 4 looks in detail at single-mode buckling phenomena at distinct critical points. For each form of instability the simplest universal expression for the local governing potential function is given: link models, struts, and frames are used as illustrative examples. The fold catastrophe is seen to generate first, the limit or snap-buckling point, and second, within a bifurcational view, the asymmetric point of bifurcation. The cusp generates the stable-symmetric and unstable-symmetric points of bifurcation: requirements of structural stability under general and bifurcational formalisms are discussed. Routes through catastrophes are used to illustrate the emergence of the bifurcational viewpoint, and a non-symmetric control route through a cusp is seen to generate a cut-off point familiar to engineers in the buckling of shallow arches and domes. Two higher-order one-mode singularities are finally discussed, with a consideration of the reduced Euler buckling load used in approximate engineering analysis.

Chapter 5 is devoted to a new analysis of stress-free and pre-stressed arches which are now used almost universally as a classical laboratory demonstration of imperfection-sensitivity in the unstable-symmetric point of bifurcation. Buckling, post-buckling, and imperfection-sensitivity predictions agree well with the available experimental results of Roorda. The explicit identification of a constraint condition throws new light on the complex contorted equilibrium paths of arches under central point load.

Chapter 6 starts by looking at the simultaneous compound buckling of a spatially guyed cantilever. The post-buckling and imperfection-sensitivity of the semi-symmetric points of bifurcation are then discussed, including the effects of four control parameters, namely the load, an imperfection in each buckling mode, and a splitting parameter (normally a geometric parameter in a particular buckling problem) that controls the gap between the two distinct bifurcations involved. As an example of the universal unfolding of the umbilic catastrophes, the homeoclinal, anticlinal, and monoclinal points of bifurcation are studied in detail, and computed imperfection-sensitivity surfaces are presented and described. Fully asymmetric points of bifurcation are classified using a Lagrange Multiplier technique, and routes through the umbilic catastrophes are explored. Higher-order two-mode singularities are sketched, including parabolic umbilic, recently identified as of fundamental significance in interactive buckling, and the double cusp, which is relevant to the compound buckling of elastic plates.

Chapter 7 presents a comprehensive bifurcation analysis of the general system governed by a potential energy function of n coordinates and h controls, first published in the *Philosophical Transactions of the Royal Society*. The system is presumed to exhibit a single-valued fundamental path under the evolution of a distinctive primary control, normally a load. Sliding coordinates are then used to define a new incremental energy function. Active and passive coordinates are segregated, and the latter are eliminated locally using a preliminary perturbation scheme to give a transformed energy function of the active coordinates alone: the original equilibrium and stability axioms hold good for this new energy function of m active coordinates. Appropriate identities for the specification of equilibrium states, critical states, and secondary bifurcations are presented explicitly, and alternative perturbation schemes are outlined. The concepts of generalized loads and generalized imperfections are introduced, and some hints for computer solutions are given. The comprehensive bifurcation analysis is illustrated by application to the semi-symmetric branching points which involve compound buckling in two simultaneous modes.

The last chapter looks at some engineering buckling problems in the light of the foregoing phenomena and analytical techniques. The elimination of passive coordinates within a diagonalized energy formulation shows explicitly how the *total* quartic energy derivatives with respect to the active coordinates are contaminated by cubic derivatives of the non-critical passive buckling modes: this is a crucial point in the following discussions. A brief look at structural optimization and its associated symmetries highlights the important role that these play in all buckling problems.

The distinctive post-buckling of struts and plates are then examined, and an approximate energy analysis due to Koiter is used for the latter to highlight the previously mentioned general features. The practically important interactive buckling of stiffened structures, very much a current research topic,[7] is examined in the light of the controlling parabolic umbilic catastrophe. The higher-order singularities governing the instability of compressed cylindrical and spherical shells are discussed following Koiter's classic contributions. The chapter ends with a brief discussion of the dead and rigid loading of laboratory model structures, while a proof of the Lagrange and Hamilton equations is given as an Appendix.

An extensive list of modern references supplements and updates the comprehensive lists in our earlier books.[1,2,7]

1

The General Conservative System

In this opening chapter we consider the dynamics and stability of a general *n*-degree-of-freedom conservative mechanical system, based on the Lagrange equations of motion. Here we define a conservative mechanical system as one whose generalized forces are completely derivable from a potential energy function $V(Q_i)$. We thus exclude gyroscopic systems, despite the fact that they do also conserve energy. We admit, however, on occasions the presence of a small amount of positive definite viscous damping. This changes pathological centres into asymptotically-stable foci in the phase-space, and allows the proof of the converse of the Lagrange energy theorem.

The Liapunov definition of stability is employed, and the equilibrium and stability axioms necessary for our later work are established. Conditions for a minimum of the total potential energy are analysed at length, and the chapter ends with a presentation of normal-mode linear vibration theory.

Discrete mechanical models are analysed throughout the chapter, and the theory is further illustrated by the strut analyses of the following chapter.

1.1 GENERALIZED COORDINATES

Spatial configurations of our general mechanical system are to be specified by a set of *n* *generalized coordinates*, written as

$$Q_1, Q_2, Q_3, \ldots Q_n$$

or more briefly as Q_i, where *i* is understood to take values from one to *n*. Here the number *n* represents the *degree of freedom* of the system. We require that there shall be a unique one-to-one correspondence between the spatial configurations of our system and the set of values of the coordinates. Thus if we introduce an *n*-dimensional *state space* by associating the algebraic variables Q_i with a set of *n* rectangular axes, as shown schematically in Figure 1.1, there will be a unique, one-to-one correspondence between spatial states of the system and the points of this space.

This one-to-one correspondence need not always be global, but must hold locally to the region of our immediate discussion. For example, local configurations of the rigid pendulum of Figure 1.2 can always be specified by the

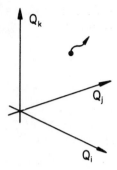

Figure 1.1 Schematic diagram of a trajectory in state space

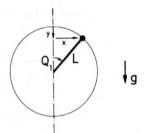

Figure 1.2 A simple rigid pendulum

angle Q_1 as a single generalized coordinate. Alternatively, the horizontal displacement x can be employed, being valid except in the region of $Q_1 = 90$ degrees or 270 degrees: as a second alternative generalized coordinate we can employ the vertical displacement y except in the region of $Q_1 = 0$ degrees or 180 degrees.

For the small-but-finite deflections of a beam or stretched wire, the mathematical continuum problem can be *discretized* by either a classical modal analysis or by a numerical finite-element analysis. In the classical analysis we might, for example, use the harmonic amplitudes Q_1, Q_2, Q_3,... as our generalized coordinates as illustrated in Figure 1.3, and the fact that the number of coordinates, n, is now strictly infinite will be largely ignored in our discussions. This step, although naturally distressing to mathematicians, rarely gives rise to any real problems in physics and engineering.

Just as any point in state space represents a unique admissible spatial position of the system, so a trajectory in state space is assumed to represent a unique admissible motion of the system. In the terminology of Synge and Griffith[8] the general mechanical system is therefore *holonomic* with, for example, no differential constraints that can arise in the three-dimensional rolling of wheels.

The general system is also assumed to have no time-dependent constraints and

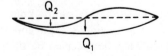

Figure 1.3 Discretization of a continuum by the use of Fourier harmonics

no time-dependent forces, so the time t will not appear explicitly in the potential or kinetic energy expressions, and it is therefore also *scleronomic*.[8]

1.2 LAGRANGE EQUATIONS

We now suppose our general mechanical system to be *conservative* and *undamped*, comprising, for example, internal elastic elements and external conservative fields (gravitational or otherwise), so that all forces of the system are derivable from a *total potential energy function V* which is a function of only the n generalized coordinates Q_i. We write this simply as

$$V = V(Q_i) \tag{1.1}$$

The kinetic energy T will be primarily a quadratic function of the rates of change of the coordinates

$$\frac{dQ_i}{dt} \equiv \dot{Q}_i \tag{1.2}$$

so we can write

$$\begin{aligned} T = &\tfrac{1}{2}T_{11}\dot{Q}_1^2 + \tfrac{1}{2}T_{12}\dot{Q}_1\dot{Q}_2 + \cdots + \tfrac{1}{2}T_{1n}\dot{Q}_1\dot{Q}_n \\ &+ \tfrac{1}{2}T_{21}\dot{Q}_2\dot{Q}_1 + \tfrac{1}{2}T_{22}\dot{Q}_2^2 + \cdots + \tfrac{1}{2}T_{2n}\dot{Q}_2\dot{Q}_n \\ &+ \tfrac{1}{2}T_{31}\dot{Q}_3\dot{Q}_1 + \cdots \\ &+ \cdots \end{aligned} \tag{1.3}$$

This can be written more compactly as

$$T = \frac{1}{2}\sum_{i=1}^{i=n}\sum_{j=1}^{j=n} T_{ij}\dot{Q}_i\dot{Q}_j \tag{1.4}$$

and even more concisely as

$$T = \tfrac{1}{2}T_{ij}\dot{Q}_i\dot{Q}_j \tag{1.5}$$

if we adopt the dummy-suffix or tensor notation due to Einstein, which says that any suffix occurring more than once in a product must be summed over all its values. In an expression of this type in which the ultimate meaningful coefficient of $\dot{Q}_5\dot{Q}_7$ (say) is $\tfrac{1}{2}(T_{57} + T_{75})$ it is convenient to specify, as we are quite free to, that the matrix T_{ij} is symmetric, so the $T_{57} = T_{75}$, etc.: the ultimate meaningful coefficient can then be written more compactly as simply T_{57}.

This form for the kinetic energy is quite general if we acknowledge that the coefficients T_{ij} may themselves be functions of the generalized coordinates (but never a function of the rates, or of the time t explicitly). Whether, in a given problem, the coefficients T_{ij} are, or are not, functions of the Q_i depends not necessarily just on the system but also on the coordinates used to describe the system, as we shall illustrate shortly for the simple rigid pendulum. To remind us of this possible dependence we write the set of coefficients T_{ij} as $T_{ij}(Q_k)$, so that our final form for the *kinetic energy function* is

$$T = \tfrac{1}{2}T_{ij}(Q_k)\dot{Q}_i\dot{Q}_j \tag{1.6}$$

Since no elemental contribution to the kinetic energy of a system can be negative, this is a positive–definite form being always positive unless the system is completely at rest (when it is zero).

If we introduce the *Lagrangian function*, $\mathscr{L}$, as simply the difference between these two energies,

$$\mathscr{L}(Q_i, \dot{Q}_j) \equiv T(Q_i, \dot{Q}_j) - V(Q_k) \qquad (1.7)$$

all dynamical motions of our general system will be governed by the Lagrange equations

$$\frac{\mathrm{d}}{\mathrm{d}t}\frac{\partial \mathscr{L}}{\partial \dot{Q}_1} - \frac{\partial \mathscr{L}}{\partial Q_1} = 0$$

$$\frac{\mathrm{d}}{\mathrm{d}t}\frac{\partial \mathscr{L}}{\partial \dot{Q}_2} - \frac{\partial \mathscr{L}}{\partial Q_2} = 0$$

$$\cdots \qquad (1.8)$$

$$\frac{\mathrm{d}}{\mathrm{d}t}\frac{\partial \mathscr{L}}{\partial \dot{Q}_n} - \frac{\partial \mathscr{L}}{\partial Q_n} = 0$$

These are written more concisely as

$$\frac{\mathrm{d}}{\mathrm{d}t}\frac{\partial \mathscr{L}}{\partial \dot{Q}_i} - \frac{\partial \mathscr{L}}{\partial Q_i} = 0 \quad (i = 1, 2, \ldots n) \qquad (1.9)$$

where the free suffix i (which is *not* repeated in any product) is understood to yield a set of equations as it takes all its values from one to n (even if this is not specifically indicated).

These Lagrange equations can be established from Newton's Laws, as outlined in the Appendix for a wider class of (non-conservative) systems, or they can themselves be regarded as fundamental.[8]

Because our system is conservative, we can finally note that the *total energy*

$$E = T + V \qquad (1.10)$$

will remain constant during any real motion of the system, corresponding to the principle of *conservation of energy*.

Example: a rigid pendulum

We shall illustrate the foregoing theory for the pendulum of Figure 1.2, comprising a light rigid rod of length L, freely pivoted at one end and carrying a concentrated mass m at its other free end. It is assumed to be acted upon by an external gravitational field of uniform strength g, as would be valid if L were small compared with the radius of the earth. This model is of our general type with a single degree of freedom, so that $n = 1$.

The total potential energy of the model, ignoring the mass of the rod, is simply the gravitational potential of the concentrated mass m, and measuring its height arbitrarily from the pivot as datum, we have

$$V(Q_1) = mgL\cos Q_1 \qquad (1.11)$$

This can if necessary be expanded as

$$V(Q_1) = mgL(1 - \tfrac{1}{2}Q_1^2 + \cdots) \qquad (1.12)$$

valid for small Q_1.

The kinetic energy is simply $\tfrac{1}{2}m$ times the square of the tip velocity (ignoring again the mass of the rod), so

$$T(\dot{Q}_1) = \tfrac{1}{2}mL^2\dot{Q}_1^2 \qquad (1.13)$$

We notice in passing that had we employed not Q_1 but the horizontal displacement x as our single generalized coordinate we would have obtained

$$V(x) = \pm mgL[1 - (x/L)^2]^{1/2} \qquad (1.14)$$

and

$$T(x, \dot{x}) = \tfrac{1}{2}mL^2(\dot{x}/L)^2[1 - (x/L)^2]^{-1} \qquad (1.15)$$

where we see that T is a function of both $\dot{x}$ and x. This is a perfectly valid formulation (away from the configurations given by $Q_1 = 90$ degrees or 270 degrees) which could be used to obtain the equations of motion in terms of the coordinate x.

We return, however, to our formulation in terms of the angle Q_1, and observing that $\mathscr{L} = T - V$ and

$$\frac{\partial \mathscr{L}}{\partial \dot{Q}_1} = \frac{\partial T}{\partial \dot{Q}_1} = mL^2\dot{Q}_1 \qquad (1.16)$$

$$\frac{\mathrm{d}}{\mathrm{d}t}\frac{\partial \mathscr{L}}{\partial \dot{Q}_1} = mL^2\ddot{Q}_1 \qquad (1.17)$$

$$\frac{\partial \mathscr{L}}{\partial Q_1} = -\frac{\partial V}{\partial Q_1} = mgL\sin Q_1 \qquad (1.18)$$

the single Lagrange equation

$$\frac{\mathrm{d}}{\mathrm{d}t}\frac{\partial \mathscr{L}}{\partial \dot{Q}_1} - \frac{\partial \mathscr{L}}{\partial Q_1} = 0 \qquad (1.19)$$

gives

$$mL^2\ddot{Q}_1 - mgL\sin Q_1 = 0 \qquad (1.20)$$

This is the exact differential equation of motion that we could have obtained by the straightforward application of Newton's Laws.

1.3 STATICAL EQUILIBRIUM

It follows from the Lagrange equations that the *necessary and sufficient* condition for statical equilibrium involving no motion of the system is the vanishing of all first derivatives of V,

$$V_i \equiv \frac{\partial V}{\partial Q_i} = 0 \quad \text{(for all } i\text{)} \tag{1.21}$$

Here, and subsequently, we are using the 'equals in all respects' symbol $\equiv$ to indicate a simple notational equivalence or definition.

Following our earlier studies[1] we write this condition formally as an axiom, to emphasize its fundamental role in the subsequent theory.

AXIOM I

A stationary value of the total potential energy with respect to the generalized coordinates is necessary and sufficient for the equilibrium of the system.

We shall see that we need only one further axiom based on the total potential energy, concerning the stability of equilibrium, to provide the foundations of a substantial body of practically important work in the theory of elastic stability.[1]

Now a general Taylor or power series expansion of the energy V about an equilibrium state $Q_i = Q_i^E$ can be written in terms of the incremental coordinates

$$q_i \equiv Q_i - Q_i^E \tag{1.22}$$

as

$$V = V^E + \sum_{i=1}^{i=n} \left.\frac{\partial V}{\partial Q_i}\right|^E q_i + \frac{1}{2}\sum_{i=1}^{i=n}\sum_{j=1}^{j=n} \left.\frac{\partial^2 V}{\partial Q_i \partial Q_j}\right|^E q_i q_j$$

$$+ \text{higher-order terms} \tag{1.23}$$

In this we can always ignore the arbitrary constant $V^E \equiv V(Q_i^E)$ and the following linear term vanishes completely by virtue of our equilibrium condition $V_i^E = 0$. The Taylor series thus starts with the quadratic form, and if we write

$$\left.\frac{\partial^2 V}{\partial Q_i \partial Q_j}\right|^E \equiv V_{ij}^E, \text{ etc.} \tag{1.24}$$

and employ the tensor summation convention, we have concisely

$$V = \tfrac{1}{2} V_{ij}^E q_i q_j + \text{higher-order terms} \tag{1.25}$$

Example: a rigid pendulum

Using by way of an illustration our derived total potential energy expression for the rigid pendulum

$$V = mgL\cos Q_1 \tag{1.26}$$

which for small values of Q_1 can be expanded about the inverted equilibrium state $Q_1 = 0$ as

$$V = mgL(1 - \tfrac{1}{2}Q_1^2 + \cdots) \tag{1.27}$$

we have

$$V_1 \equiv \frac{\partial V}{\partial Q_1} = -mgL\sin Q_1 \tag{1.28}$$

which vanishes at the two obvious equilibrium positions given by

$$Q_1 = 0 \text{ or } 180 \text{ degrees} \tag{1.29}$$

These statical equilibrium states can alternatively be derived by setting $\partial V/\partial x = 0$, since the horizontal displacement x is a valid coordinate in the vicinity of the two equilibrium states. Notice, however, that had we used the vertical displacement y, which is *invalid* at these two points, we would have obtained

$$V = mg(L - y) \tag{1.30}$$

giving us the first derivative

$$\partial V/\partial y = -mg \tag{1.31}$$

which can never be zero.

1.4 STABILITY DEFINITION

Stability and instability of a statical equilibrium state are conventionally defined in terms of the *free motions* of the system following an *infinitesimal* and *once-and-for-all* disturbance from the equilibrium state. An excellent and most readable account of this is given in the book by La Salle and Lefschetz.[9]

This classical definition is, to an engineer or physicist, completely unrealistic, since a real system will inevitably be subjected *continuously* throughout its life to small but *finite*, essentially random, disturbances arising, for example, from wind pressures, foundation vibrations, and even molecular thermal motions. The only justification for the definition is its extreme mathematical simplicity and tractability. It does, however, yield correct answers in a wide range of practical situations, pointing to the fact that the disturbances are often comparatively small, and that the once-and-for-all definition is often synonymous with more realistic and less tractable statistical definitions.

The free motions under discussion are best viewed in the $2n$-dimensional *phase-space* associated with the coordinates and their derivatives, Q_i and $\dot{Q}_j$. All possible free trajectories of the moving system fill this space and only cross or intersect at statical equilibrium states. They thus correspond to a *vector field* in the phase space which is best analysed in general terms by the Hamiltonian form of the Lagrange equations.[8,10] (see appendix).

The *Liapunov definition* of stability can thus be informally written as:

If all dynamical motions in phase-space in the vicinity of a statical equilibrium state remain in the vicinity for all time, then the state is termed stable. If, in addition, all such motions tend to the equilibrium state as time increases indefinitely, the state is termed asymptotically stable.

If any one dynamical motion in phase-space starting in the vicinity of a statical equilibrium state carries the system away from the vicinity, then the state is termed unstable.

We have included in these definitions the concept of *asymptotic stability*, so that we can discuss conservative systems to which a little damping has been added.

Example: a rigid pendulum

The quickest and simplest way to obtain the phase trajectories of a one-degree-of-freedom conservative system, such as our rigid pendulum, is not to use the Lagrange equations but to invoke the principle of conservation of energy by setting

$$E(Q_1, \dot{Q}_1) \equiv T + V = K \tag{1.32}$$

where K is a constant of a particular motion. This yields for the pendulum

$$\tfrac{1}{2} mL^2 \dot{Q}_1^2 + mgL \cos Q_1 = K \tag{1.33}$$

which gives the trajectories of Figure 1.4 as K takes different fixed values corresponding to different starting conditions. These trajectories can, in fact, be viewed simply as contours of constant total energy E. The direction of the arrows on these curves is readily obtained by inspection since, with $\dot{Q}_1$ positive, Q_1 must be increasing.

The trajectories represent every possible dynamical motion of the pendulum, the picture being cyclic since $Q_1 = a$ and $Q_1 = a + 360$ degrees represent identical spatial states of the system. In fact, it would be topologically neater to draw the phase pictures on a cylinder, with $\dot{Q}_1$ along the axis and Q_1 around the circumference.[2]

We see that we have a minimum of the *total* energy $E(Q_1, \dot{Q}_1)$ at the point $Q_1 = 180$ degrees, $\dot{Q}_1 = 0$, the circling trajectories corresponding to the small-amplitude free vibrations of the pendulum about its natural hanging configuration. No dynamical paths leave the vicinity of this equilibrium state, confirming that the hanging configuration is indeed stable according to our earlier definition. The state is not, however, asymptotically stable because of the lack of damping in our idealized pendulum.

We see next that we have a saddle-point of $E(Q_1, \dot{Q}_1)$ at the point $Q_1 = \dot{Q}_1 = 0$ (corresponding to a minimum of T but a maximum of V), the trajectories indicating diverging motions of the pendulum away from the inverted con-

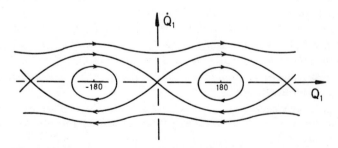

Figure 1.4 The phase-space trajectories of a rigid pendulum

figuration. Dynamical paths leave the vicinity of this equilibrium state, confirming that the inverted pendulum is unstable by our earlier definition.

The critical trajectories that enter and leave this unstable state correspond to dynamical motions involving infinite time, as the speed of the representative point in phase-space is tending to zero near this state.

The free motion followed in practice by a given undamped pendulum depends, of course, on the starting conditions, that is, the initial values of the coordinate and its time derivative, which serve to define a unique trajectory. The continuous unbounded motions occurring at large positive and negative values of $\dot{Q}_1$ represent the rapid complete rotations of the pendulum that could be induced by a high-velocity start.

1.5 ENERGY THEOREMS

For conservative mechanical systems, we are fortunate in having powerful energy theorems that allow stability conclusions to be drawn simply from an inspection of the total potential energy function $V(Q_i)$.

Consider a statical equilibrium state of a system for which the total potential energy function has a complete relative minimum, so that in state-space energy contours are closed surfaces around the equilibrium state. The energy function V will then have the shape of a bowl in $n+1$ dimensions, as illustrated schematically in Figure 1.5, in which we show the closed contour for $V = k$.

Clearly, any dynamical motion of the system, starting near the equilibrium state with *total* energy E less than k, can never escape from the closed-contour $V = k$ by virtue of the principle of conservation of energy. The equilibrium state is therefore stable and we have the theorem of Lagrange:

An equilibrium state at which the total potential energy is an isolated local minimum is necessarily stable.

The specification of a *local* energy minimum means that the ball of Figure 1.6 would be deemed stable, despite the fact that a deeper total potential energy minimum occurs elsewhere. To draw attention to the existence of a known deeper energy trough we sometimes use the term *metastable* to describe this circumstance.

If we finally consider the addition to our general conservative system of a little

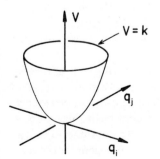

Figure 1.5 A schematic $(n+1)$ dimensional bowl for $V(Q_i)$

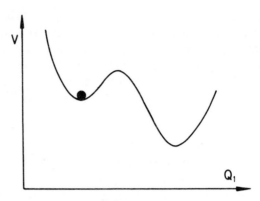

Figure 1.6 A ball rolling on a curve, illustrating
the concept of metastability

positive definite damping that absorbs energy whenever the system moves, it is
clear that the stability predicted by Lagrange's theorem will be changed from
non-asymptotic to asymptotic.

The converse of the Lagrange theorem cannot, surprisingly, be proved in
complete generality for our conservative mechanical system without appealing to
the existence of a little damping, and we would refer to the comprehensive paper
of Koiter[11] for a modern appraisal of this situation. We might add, however, that
it is widely assumed that a complete proof will eventually become available.

We then proceed to outline the proof of this converse theorem with an appeal
to a little positive–definite damping. We consider an equilibrium state at which
$V(Q_i)$ is not a local minimum, being therefore either a local maximum or a local
multi-dimensional saddle point. There will then exist, close to this equilibrium
state, configurations of the system that have a lower value of the total potential
energy V. Now a motion starting at rest with zero kinetic energy from such a
displaced (non-equilibrium) state can never return to the fundamental equilibrium state even without damping: and with positive–definite damping the total
energy $E = T + V$ must steadily decrease. Moreover, since we are assuming the
fundamental equilibrium state to be isolated, with no other equilibrium state in
the immediate vicinity, the system cannot come to rest locally, and so with
damping must eventually leave the neighbourhood. We therefore have instability, and we can write the converse of the Lagrange theorem:

*An equilibrium state at which the total potential energy is not an isolated local
minimum is necessarily unstable.*

Now all real mechanical systems in physics and engineering will have *some*
positive–definite damping since all motion involves the dissipation of some
energy as heat. Therefore, even when *nominally* considering an undamped system,
we shall, from a practical point of view, be able to accept both the Lagrange
theorem and its converse. Following our earlier text,[1] we combine these two
theorems into what we might regard as a basic *axiom* of our general theory.

AXIOM II

A complete relative minimum of the total potential energy with respect to the generalized coordinates is necessary and sufficient for the stability of an equilibrium state of the system.

This, together with our earlier Axiom I specifying a stationary total potential energy for equilibrium, can be viewed as an essential foundation for the analysis of discrete conservative systems.

It follows that the precise form of the *kinetic* energy function will rarely be of any concern in establishing the stability or instability of a system. We can thus forget about it completely, but before doing so we shall present here a brief discussion of the situation. That the form of the kinetic energy has no effect on the statical equilibrium states of the system is quite expected, but since stability is a dynamical concept concerning motions of the system we might express some surprise that a change in the disposition of inertial mass can effect neither a stabilization nor a destabilization of a system. That this is indeed the case has quite an important consequence in our treatment of structural loadings.

It is customary in structural engineering to glibly postulate the existence of a dead vertical load P, assumed to retain its magnitude and direction as the systems deflect, and it is pertinent to inquire how such a load might arise in practice. In such cases, a civil engineer would usually have in mind a mass m hanging in a uniform gravitational field g, where mg is equal to P. This is fine under static conditions, but we see that under dynamic conditions this gravity loading does not quite fit the bill: during motion of the system the hanging mass will experience vertical accelerations, so the tension in the wire will not be equal to $mg = P$.

In the variational mechanics formulation the two structural systems of Figure 1.7 will have identical total potential energy functions comprising the stored

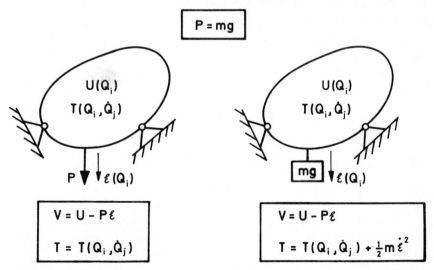

Figure 1.7 Two structural systems with different dynamic responses but identical static responses

elastic strain energy $U(Q_i)$ and the gravitational potential of the load $-P\mathscr{E}(Q_i)$, where $\mathscr{E}(Q_i)$ is the vertical displacement of the mass, but the kinetic energy functions will differ by an additive factor

$$\tfrac{1}{2}m\dot{\mathscr{E}}^2 \tag{1.34}$$

and herein lies the key. By virtue of our two axioms the two systems will have the *same response* as far as their equilibrium and stability are concerned, and so the distinction between the systems will often be of no concern.

Suffice it to say that in structural loading situations of any complexity it is usually much safer and more elegant to enlarge the system under consideration to embrace the loading feature, and then to associate a 'loading parameter' with a fundamental parameter of the enlarged system such as a mass or even a gravitational constant: we prefer the hanging mass to the nebulous force P. By doing this we can be sure that we are indeed dealing with a conservative system: non-conservative systems, such as arise under fluid loading,[2] are out of the scope of the present study.

We can finally see this stability discussion at work for our rigid pendulum by forming

$$\frac{\partial^2 V}{\partial Q_1^2} \equiv V_{11} = -mgL\cos Q_1 \tag{1.35}$$

This second energy derivative is positive indicating a stable minimum for the hanging pendulum with $Q_1 = 180$ degrees but negative indicating an unstable maximum for the inverted pendulum with $Q_1 = 0$.

1.6 CONDITIONS FOR A MINIMUM

To establish the stability or instability of an equilibrium state it simply remains to inspect our total potential energy function $V(Q_i)$ in the vicinity of this state: if V is a local minimum we have stability; otherwise we have instability.

We have seen that a general Taylor or power series expansion about an equilibrium state will have the form

$$V = \tfrac{1}{2}V^E_{ij}q_iq_j + \tfrac{1}{6}V^E_{ijk}q_iq_jq_k + \cdots \tag{1.36}$$

where

$$V^E_{ijk} \equiv \left.\frac{\partial^3 V}{\partial Q_i\,\partial Q_j\,\partial Q_k}\right|^E \quad \text{etc.} \tag{1.37}$$

and all subscripts are summed from one to n. Since locally the incremental coordinates q_i, as defined in equation (1.22), will be small, the terms of this series have decreasing magnitudes, and we must clearly *first* inspect the dominant quadratic form

$$\delta^2 V = \tfrac{1}{2}V^E_{ij}q_iq_j \tag{1.38}$$

which we can write out in full as

$$\delta^2 V = \tfrac{1}{2}V^E_{11}q_1^2 + \tfrac{1}{2}V^E_{12}q_1q_2 + \cdots + \tfrac{1}{2}V^E_{1n}q_1q_n$$
$$+ \tfrac{1}{2}V^E_{21}q_2q_1 + \tfrac{1}{2}V_{22}q_2^2 + \cdots + \tfrac{1}{2}V^E_{2n}q_2q_n$$
$$+ \tfrac{1}{2}V^E_{31}q_3q_1 + \cdots$$
$$+ \cdots$$
$$+ \tfrac{1}{2}V^E_{n1}q_nq_1 + \tfrac{1}{2}V^E_{n2}q_nq_2 + \cdots + \tfrac{1}{2}V^E_{nn}q_n^2 \tag{1.39}$$

Here we see squared terms such as $\tfrac{1}{2}V^E_{22}q_2^2$ and also cross-terms such as $\tfrac{1}{2}V^E_{12}q_1q_2$, and it is our intention to eliminate the latter by a *diagonalization* procedure. To do this we introduce the non-singular, reversible linear transformation

$$q_i = \alpha_{ij}u_j \quad \text{with} \quad |\alpha_{ij}| \neq 0 \tag{1.40}$$

which can be written out in full as

$$q_1 = \alpha_{11}u_1 + \alpha_{12}u_2 + \cdots + \alpha_{1n}u_n$$
$$q_2 = \alpha_{21}u_1 + \alpha_{22}u_2 + \cdots + \alpha_{2n}u_n$$
$$q_3 = \alpha_{31}u_1 + \cdots$$
$$\cdots$$
$$q_n = \alpha_{n1}u_1 + \alpha_{n2}u_2 + \cdots + \alpha_{nn}u_n \tag{1.41}$$

Substituting this into the second variation $\delta^2 V$ gives us the quadratic form in terms of the new principal incremental coordinates u_i, and it is a well-known mathematical result[1] that we can always find, in a particular problem, values of the α_{ij} that ensure that in the new coordinates we have simply a sum of squares, which we write as

$$\delta^2 V = \tfrac{1}{2}C_1u_1^2 + \tfrac{1}{2}C_2u_2^2 + \cdots + \tfrac{1}{2}C_nu_n^2 \tag{1.42}$$

This diagonalization has eliminated the cross-terms of $\delta^2 V$, leaving us with just diagonal coefficients

$$\left.\frac{\partial^2 V}{\partial u_i^2}\right|^E \equiv C_i \tag{1.43}$$

The new coordinates u_i will be called *principal coordinates* and the coefficients C_i will be termed *stability coefficients*.

In visualizing this and subsequent transformations, it is best to imagine a *fixed* energy surface (representing V or $\delta^2 V$), simply described by a *new* set of coordinates, as illustrated in Figure 1.8. This emphasizes that we can make invariant statements about the forms of V or $\delta^2 V$ which do not depend on the coordinates by which they might be described.

The new coordinates u_i are, of course, *independent* of one another but may be associated with *oblique* axes in rectangular q_i space, as illustrated schematically in Figure 1.8.

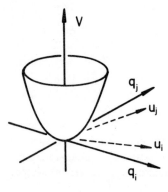

Figure 1.8 An invariant energy bowl under coordinate transformations

Let us now denote the lowest of the stability coefficients by C_i^*. Then if C_i^* is positive, all the stability coefficients are positive, and the second variation $\delta^2 V$ is clearly always positive (apart from its zero value at the origin). We can then make the invariant statement that $\delta^2 V$ is *positive–definite*, and since non-zero values of $\delta^2 V$ will locally dominate the full Taylor expansion, it follows that the complete energy function V will be a relative minimum. Such an equilibrium state, with a minimum guaranteed by a positive–definite second variation, is termed *thoroughly stable*.

If C_i^* is negative, then for a corresponding displacement from equilibrium, $\delta^2 V$ will be negative. We then say, invariantly, that the quadratic form *admits negative values*. These negative values cannot be destroyed by higher terms in the Taylor expansion, so V is not now a minimum and we say that the equilibrium state is *thoroughly unstable*.

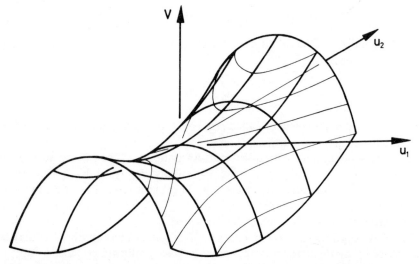

Figure 1.9 A total potential energy surface showing stability with respect to u_2 but instability with respect to u_1. Such a surface is called a saddle point

If C_i^* is zero, the second variation $\delta^2 V$ is then identically zero in the corresponding direction, and we say that it is *positive–semi-definite*. In this direction higher-order terms of the Taylor series will clearly play a decisive role, so the quadratic form now supplies no complete conclusion. A higher-order inspection of V must then be made, as we shall outline in the following section.

Any equilibrium state with one or more zero stability coefficients is called *critical*, and, depending on the result of further investigation, it will be termed *critically stable* or *critically unstable*. If only one C_i is zero we shall say that we have a *distinct* critical point, while if more than one C_i is zero we shall say that we have a *compound* critical point.

We can observe that the *stability determinants* at E,

$$\left| \frac{\partial^2 V}{\partial Q_i \partial Q_j} \right| \equiv \left| \frac{\partial^2 V}{\partial q_i \partial q_j} \right| \tag{1.44}$$

and

$$\left| \frac{\partial^2 V}{\partial u_i \partial u_j} \right| \equiv C_1 C_2 C_3 \ldots C_n \tag{1.45}$$

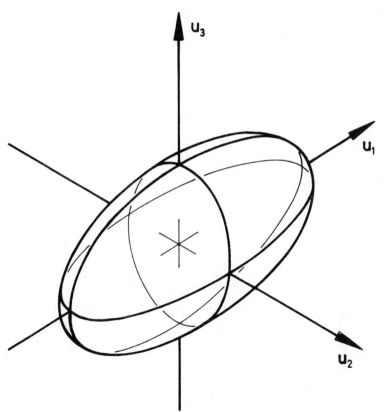

Figure 1.10 Contour of constant total potential energy about a stable equilibrium state

16

are related by the equation

$$|\alpha_{ij}|^2 \cdot \left|\frac{\partial^2 V}{\partial q_i \partial q_j}\right| = \left|\frac{\partial^2 V}{\partial u_i \partial u_j}\right| \qquad (1.46)$$

so *both* of these determinants will vanish at a critical equilibrium state. This vanishing of the original determinant $|\partial^2 V/\partial q_i \partial q_j|$ allows the determination of critical loads without resort to a scheme of diagonalization, as we shall see in the following example.

For a more complete discussion of these stability conditions we can refer to our earlier text.[1]

In discussions of the present type, there are two useful methods of visualization which we now use to gain some familiarity with common energy configurations. First, for two degrees of freedom, it is possible to envisage, as we have done before, a V-surface in the $(n + 1)$ dimensional $V - u_i$ space, as shown in Figure 1.9; this is clearly drawn in the region of an equilibrium state which is unstable with respect to one principal coordinate u_1 and stable with respect to the second principal coordinate u_2, and thus takes the form of a saddle point. Second, for three degrees of freedom we can draw V-contours in u_i space, as shown in Figure 1.10 and 1.11; these show stable and unstable equilibrium states, respectively. For the stable

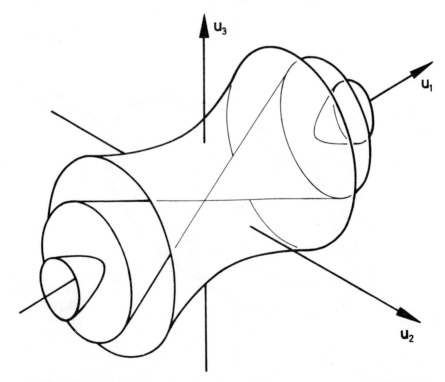

Figure 1.11 Contours of constant total potential energy showing stability with respect to u_2 and u_3 but instability with respect to u_1

state of Figure 1.10, the energy always increases as we emerge from the equilibrium state, while in Figure 1.11 we see that the energy decreases as we emerge from the equilibrium state in the direction u_1 but rises as we proceed in the u_2 or u_3 directions. The conical surface in this figure represents states that have the same energy level as the equilibrium state, states within the cone having a lower energy level and states outside the cone having higher energy.

Example: a two-hinged strut

We shall illustrate the foregoing theory by considering the behaviour of the two-hinged strut shown in a *deflected* state in Figure 1.12.

Here three light rigid rods each of length L are hinged together to form a chain of length $3L$. One end of the chain is pivoted at a fixed point while the other is free to move axially towards this pivot. Relative rotation of the rods is resisted by two rotational springs each of stiffness k at the internal joints. These springs are unstrained when the links lie in a straight line, so that with no applied load the links lie in a straight horizontal line between the supports. The system is loaded by a dead compressive load P which is assumed to retain its original magnitude and direction as the links deflect.

This system can be thought of as a finite-dimensional model of the *continuous* compressed elastic column that we shall examine in the next chapter.

The system has two degrees of freedom, and the vertical deflections of the internal joints are denoted by $Q_1 L$ and $Q_2 L$ as shown, so the total deflected form of the system may be fully represented by specifying the non-dimensional generalized coordinates Q_1 and Q_2. These are both equal to zero in the natural unstrained shape of the column.

The strain energy stored in the two rotational springs is

$$U = \tfrac{1}{2}k\theta_1^2 + \tfrac{1}{2}k\theta_2^2 \tag{1.47}$$

which can be written as

$$U = \tfrac{1}{2}k[\sin^{-1}Q_1 - \sin^{-1}(Q_2 - Q_1)]^2$$
$$+ \tfrac{1}{2}k[\sin^{-1}Q_2 + \sin^{-1}(Q_2 - Q_1)]^2 \tag{1.48}$$

and expanding each of the trigonometric terms as a power series we have

$$U = \tfrac{1}{2}k[5Q_1^2 - 8Q_1 Q_2 + 5Q_2^2 + \text{higher-order terms}] \tag{1.49}$$

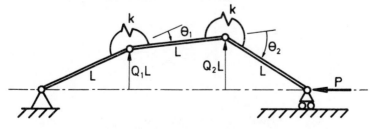

Figure 1.12 The deflected state of a two-hinged model strut

The end-shortening of the column, due to the sideways deflection is

$$\mathscr{E} = L\{3 - (1 - Q_1^2)^{1/2} - (1 - Q_2^2)^{1/2} - [1 - (Q_2 - Q_1)^2]^{1/2}\} \qquad (1.50)$$

and again expanding the terms as power series we have

$$\mathscr{E} = L[Q_1^2 - Q_1 Q_2 + Q_2^2 + \text{higher-order terms}] \qquad (1.51)$$

Now the total potential energy of this system is simply

$$V(Q_i) = U(Q_i) - P\mathscr{E}(Q_i) \qquad (1.52)$$

the latter term representing the potential energy of the applied load, so the second variation of V is simply

$$\delta^2 V = \tfrac{1}{2}k[5Q_1^2 - 8Q_1 Q_2 + 5Q_2^2] \\ - PL[Q_1^2 - Q_1 Q_2 + Q_2^2] \qquad (1.53)$$

This is precisely the quadratic form that we must inspect if we wish to examine the stability of the straight configuration, which is confirmed as an equilibrium state by the absence in the expansions of the first variation δV.

Now if we have two quadratic forms, and one of them is positive–definite (as is the quadratic form of the strain energy multiplying $\tfrac{1}{2}k$, since it is derived from a sum of two squares) it is always possible to find a linear change of variables to *simultaneously* diagonalize the two forms. This is, in fact, achieved in this particular problem by the non-singular transformation

$$u_1 = \frac{Q_1 + Q_2}{2} \quad u_2 = \frac{Q_1 - Q_2}{2} \qquad (1.54)$$

which has the inverse

$$Q_1 = u_1 + u_2, \quad Q_2 = u_1 - u_2 \qquad (1.55)$$

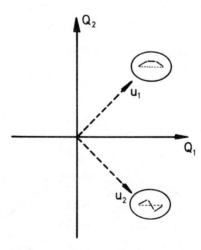

Figure 1.13 The principal coordinates of the model strut in Q_i space

To locate the new u_i axes in Q_j space we note here that $u_1 = 0$ gives $Q_1 = -Q_2$ while $u_2 = 0$ gives $Q_1 = Q_2$. So the u_i axes are, in this instance, rectangular and at 45 degrees relative to the Q_j axes, as shown in Figure 1.13.

In terms of these *principal* u_i coordinates we find by direct substitution

$$\delta^2 V = \tfrac{1}{2}k[2u_1^2 + 18u_2^2]$$
$$- PL[u_1^2 + 3u_2^2] \tag{1.56}$$

and comparing this sum of squares with our standard form

$$\delta^2 V = \tfrac{1}{2}C_1 u_1^2 + \tfrac{1}{2}C_2 u_2^2 \tag{1.57}$$

we find the *stability coefficients*

$$C_1 = 2k - 2PL$$
$$C_2 = 18k - 6PL \tag{1.58}$$

If the applied load is zero, the straight configuration is naturally stable with

$$C_1 = 2k = \text{positive}$$
$$C_2 = 18k = \text{positive} \tag{1.59}$$

and the energy function has a local minimum, as shown in the bottom right-hand diagram of Figure 1.14.

As the applied axial compressive load P is increased, the two stability coefficients decrease linearly with P until C_1 reaches zero at the first critical load P^1 given by

$$P^1 = k/L \tag{1.60}$$

At this load the curvature of the energy surface has dropped to zero in the u_1 direction, so that the second variation yields the cylindrical form shown on the right of Figure 1.14. We say the column *buckles* in *mode* u_1 at this critical load.

Beyond P^1 we see that C_1 is negative while C_2 is as yet still positive. The surface of the second variation of V now has the form of a saddle-point, curving up in the u_2 direction but down in the u_1 direction, as shown. The fundamental straight configuration of the model is now unstable. To supply more information, we could, if we wished, say that it is unstable *with respect to* u_1 but stable *with respect to* u_2.

As we continue to increase the applied load P, the second stability coefficient C_2 finally reaches zero at the second critical load P^2 given by

$$P^2 = 3k/L \tag{1.61}$$

At this second buckling load $\delta^2 V$ is again cylindrical, as illustrated, and for even higher values of load it is a local maximum. We could say that our model strut buckles in mode u_2 at the second critical load P^2, although this condition could not be achieved experimentally without the addition of a constraint to inhibit the earlier buckling into mode u_1.

We emphasize that each energy surface drawn on the right-hand side of Figure 1.14 relates to a given fixed value of the applied load P.

20

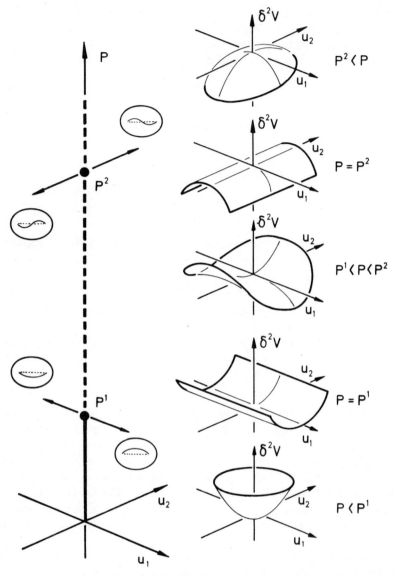

Figure 1.14 The progressive loss of stability of the model strut as the compressive load P is increased

The mode-forms we mentioned are easily inspected. When we speak of mode u_1 we mean the deformation that can occur with $u_2 = 0$. Now $u_2 = 0$ implies by the transformation equation that $Q_1 = Q_2$, so mode u_1 is a symmetric deformation approximating to the first harmonic of a continuous column, as shown in Figure 1.15. Similarly, $u_1 = 0$ implies $Q_1 = -Q_2$, so mode u_2 is a skew-symmetric deformation approximating to the second harmonic of a continuous column, as shown.

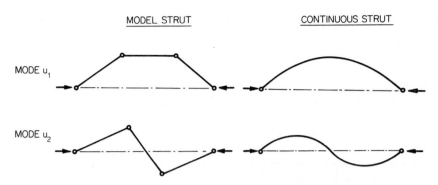

Figure 1.15 The buckling mode forms of the model and a continuous strut or column

We notice finally that the two critical loads, P^1 and P^2, could have been determined without resort to the diagonalizing transformation by equating to zero the stability determinant

$$\left|\frac{\partial^2 V}{\partial Q_1 \partial Q_2}\right| = \begin{vmatrix} (5k - 2PL) & (-4k + PL) \\ (-4k + PL) & (5k - 2PL) \end{vmatrix} = 0 \tag{1.62}$$

giving

$$9k^2 - 12kPL + 3P^2L^2 = 0 \tag{1.63}$$

This quadratic equation in PL/k can be factored as

$$\left(3 - 3\frac{PL}{k}\right)\left(3 - \frac{PL}{k}\right) = 0 \tag{1.64}$$

giving

$$\frac{PL}{k} = 1 \text{ or } 3 \tag{1.65}$$

An analysis such as this that simply aims to determine critical loads by examining the second variation of V is called a *linear eigenvalue analysis*. It supplies no information about the behaviour of the system *after* initial buckling, which is controlled by higher variations of V about the critical equilibrium state, as we shall now see.

1.7 STABILITY OF A CRITICAL STATE

The bifurcational behaviour of a system in the vicinity of a critical equilibrium state depends crucially on the stability or instability of that state.[1,12,13] For this reason we must now examine the non-linear problem of establishing the existence of a local minimum of V at a critical equilibrium state.

Suppose we are concerned to examine the form of V at a *distinct* critical point associated with the vanishing of the first stability coefficient C_1: the other stability coefficients are assumed to be positive. Then in our principal coordinates u_i our

22

Taylor expansion of V about the critical equilibrium state can be written as

$$V = \tfrac{1}{2}0u_1^2 + \tfrac{1}{2}C_2u_2^2 + \tfrac{1}{2}C_3u_3^2 + \cdots + \tfrac{1}{2}C_nu_n^2$$
$$+ \tfrac{1}{6}V^C_{ijk}u_iu_ju_k + \tfrac{1}{24}V^C_{ijkl}u_iu_ju_ku_l + \cdots \qquad (1.66)$$

where a subscript on V now denotes partial differentiation with respect to the *principal* coordinates u_i, and the C denotes evaluation at a critical equilibrium state.

As we make a trial displacement in state-space away from the critical equilibrium state $u_i = 0$ it is clear that the total potential energy will increase, since the quadratic terms dominate, in every direction except along the u_1 axis, where $\delta^2 V$ predicts no change in energy. The dominant term in this direction is V^C_{111} and if this is *non-zero* (positive or negative) it is clear that we do *not* have a local minimum.

If, however, V^C_{111} is zero, we have to look next at the quartic terms of our Taylor expansion. If V^C_{1111} is *negative* we clearly do *not* have a minimum, since V

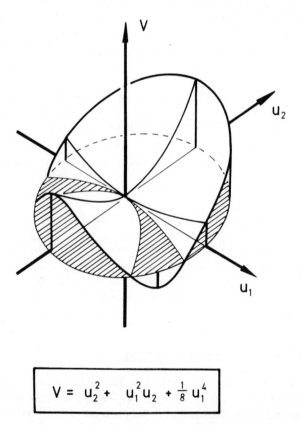

$$V = u_2^2 + u_1^2 u_2 + \tfrac{1}{8}u_1^4$$

Figure 1.16 An energy form that rises along every straight ray from the origin, but falls along some curved rays from the origin

will fall continuously away from the equilibrium state as we proceed directly along the u_1 axis. In dealing with the quartic variation, however, a new subtlety arises, because a positive value of V^C_{1111} *does not guarantee* a minimum. Such a positive value of V^C_{1111} would certainly ensure an increasing V along the u_1 axis, and hence an increasing V along *any straight ray* from the origin $u_i = 0$, but the subtlety is that it may fall along a *curved path* that leaves the origin initially in the direction of u_1. This is illustrated in Figure 1.16. In fact, for our distinct critical point with $C_1 = 0$ analysed in the diagonal coordinate system u_i, a perturbation analysis (reference 1, page 96) shows that the significant coefficient is not simply V^C_{1111} but

$$\tilde{V}^C_{1111} = V^C_{1111} - 3 \sum_{s=2}^{s=n} (V^C_{s11})^2 / V^C_{ss} \tag{1.67}$$

which we could call an intrinsic fourth derivative suffering a *contamination* from the *passive* modes u_s. We shall see more of this derivative later, in the elimination of passive coordinates (see Figure 8.1, for example).

If this intrinsic derivative is positive, then we have a local energy minimum, and any bifurcation would be of the stable-symmetric type, while if it is negative we have no local minimum, and any bifurcation would be of the unstable-symmetric type. If it is zero, the quartic terms supply no decision about stability, and we are forced to examine even higher terms in the Taylor expansion.

In the Euler strut analysis of Chapter 4 we see that the critical equilibrium state is stable by virtue of a positive V^C_{1111}, the passive contamination of equation (1.67) being entirely absent; this is due to the fundamental symmetry of the strut, as discussed further in Section 8.3. On the other hand, we also find in Chapter 4 a non-trivial contamination, and the use of the intrinsic fourth derivative, in the plate and shell model.

1.8 LINEAR VIBRATIONS

We conclude this chapter with a discussion of small-amplitude linear vibrations about a non-critical equilibrium state. About such a state a valid local truncation of the total potential energy function is

$$V = \tfrac{1}{2} V^E_{ij} q_i q_j \tag{1.68}$$

as we have seen in our discussion of maxima and minima. Now the general form of our kinetic energy function is

$$T = \tfrac{1}{2} T_{ij}(q_k) \dot{q}_i \dot{q}_j \tag{1.69}$$

and for a linear study we can eliminate the dependency on q_k by setting $q_k = 0$ in the coefficients. That is, the coefficients will be assumed to retain their equilibrium values, and we can write

$$T = \tfrac{1}{2} T^E_{ij} \dot{q}_i \dot{q}_j \tag{1.70}$$

where $T^E_{ij} \equiv T_{ij}(0)$.

Thus our Lagrangian function $\mathscr{L} = T - V$ for small vibrations about a non-critical equilibrium state is

$$\mathscr{L} = \tfrac{1}{2}T_{ij}^E\dot{q}_i\dot{q}_j - \tfrac{1}{2}V_{ij}^E q_i q_j \tag{1.71}$$

and so the Lagrange equations

$$\frac{\mathrm{d}}{\mathrm{d}t}\frac{\partial\mathscr{L}}{\partial\dot{q}_i} - \frac{\partial\mathscr{L}}{\partial q_i} = 0 \tag{1.72}$$

give us

$$T_{ij}^E\ddot{q}_j + V_{ij}^E q_j = 0 \tag{1.73}$$

This is a set of n-coupled differential equations generated as i takes the values one to n, and we see that according to our tensor dummy-suffix notation each term involves a summation over j. In deriving them we have made use of the simple rule that

$$\frac{\partial}{\partial q_s}\tfrac{1}{2}V_{ij}^E q_i q_j = V_{sj}^E q_j, \text{ etc.} \tag{1.74}$$

which can be confirmed by writing the summations out in full.

Now, as in our discussion of $\delta^2 V$, we can introduce here a non-singular linear change of coordinates

$$q_i = \alpha_{ij}u_j \tag{1.75}$$

to *simultaneously diagonalize* T_{ij}^E and V_{ij}^E. It is, in fact, always possible to simultaneously diagonalize two quadratic forms provided that one of them is positive–definite, and for a mechanical system we know the kinetic energy to be always positive–definite.[8]

In the new coordinates, we can then write

$$T_{ii}^E\ddot{u}_i + V_{ii}^E u_i = 0 \quad \text{(no summation)} \tag{1.76}$$

where the subscripts now denote differentiation with respect to the u_i, and we see that the equations are now *de-coupled*. In fact, we have the equations of n uncoupled linear oscillators, which vibrate at their own particular frequency completely independently of one another. With $V_{ii}^E = C_i$ assumed positive, the solution for the ith *normal mode of vibration* is simply

$$u_i = a_i \sin(\omega_i t + p_i) \tag{1.77}$$

where the two arbitrary constants a_i and p_i are determined by the starting conditions in this mode. The circular frequency of vibration ω_i is given by

$$\omega_i^2 = \frac{V_{ii}^E}{T_{ii}^E} \tag{1.78}$$

and we observe that in the $(u_i, \dot{u}_i)$ phase sub-space we have the centres of our single-mode discussions.

These *linear eigenvalue* solutions are seen to confirm our earlier stability discussions: they predict unbounded divergent solutions for any mode with a negative stability coefficient.

To predict the natural frequencies without resort to a scheme of diagonalization, which is not always easy to find, we can simply substitute a trial solution

$$q_j = A_j \sin \omega t \tag{1.79}$$

into the undiagonalized equations of motion

$$T^E_{ij}\ddot{q}_j + V^E_{ij}q_j = 0 \tag{1.80}$$

to obtain

$$V^E_{ij}A_j - \omega^2 T^E_{ij}A_j = 0 \tag{1.81}$$

These algebraic equations have a non-trivial solution if the determinant vanishes, so we have the *characteristic equation* for ω

$$|V^E_{ij} - \omega^2 T^E_{ij}| = 0 \tag{1.82}$$

which can be solved for the n required values of ω, as we shall see in the following example.

Example: the two-hinged strut

We return to our two-hinged strut of Figure 1.12 for which we have already determined the total potential energy matrix in the Q_i and u_i coordinates,

$$\left[\frac{\partial^2 V}{\partial Q_i \partial Q_j}\right] = \begin{bmatrix} (5k - 2PL) & (-4k + PL) \\ (-4k + PL) & (5k - 2PL) \end{bmatrix} \tag{1.83}$$

$$\left[\frac{\partial^2 V}{\partial u_i \partial u_j}\right] = \begin{bmatrix} (2k - 2PL) & 0 \\ 0 & (18k - 6PL) \end{bmatrix} \tag{1.84}$$

Now for a 'general' link of mass per unit length m with displacement a at one end and b at the other the linear kinetic energy is obtained by integration as

$$T = \tfrac{1}{2}m \int_0^L [\dot{a}(1 - x/L) + \dot{b}(x/L)]^2 dx$$

$$= \tfrac{1}{2}m[\dot{a}^2 L(1/3) + \dot{b}^2 L(1/3) + 2\dot{a}\dot{b}L(1/6)] \tag{1.85}$$

Then choosing the appropriate identities for a and b and adding the contributions from the three links we find the total kinetic energy of the system

$$T = \tfrac{1}{2}mL^2 \tfrac{2}{3}(\dot{Q}_1^2 + \dot{Q}_2^2 + \tfrac{1}{2}\dot{Q}_1\dot{Q}_2) \tag{1.86}$$

which can be written in terms of the u_i coordinates as

$$T = \tfrac{1}{2}mL^2(\tfrac{5}{3}\dot{u}_1^2 + \dot{u}_2^2) \tag{1.87}$$

We see that as well as diagonalizing the strain energy and the potential energy of the axial load, the new coordinates have also given us a diagonalized T. They are therefore the buckling modes *and* the normal vibration modes, and we have

$$\left[\frac{\partial^2 T}{\partial \dot{Q}_i \partial \dot{Q}_j}\right] = \begin{bmatrix} \frac{2}{3}mL^3 & \frac{1}{6}mL^3 \\ \frac{1}{6}mL^3 & \frac{2}{3}mL^3 \end{bmatrix} \tag{1.88}$$

$$\left[\frac{\partial^2 T}{\partial \dot{u}_i \partial \dot{u}_j}\right] = \begin{bmatrix} \frac{5}{3}mL^3 & 0 \\ 0 & mL^3 \end{bmatrix} \tag{1.89}$$

Working with the normal modes we have immediately

$$\omega_1^2 = \frac{V_{11}^E}{T_{11}^E} = \frac{6k - 6PL}{5mL^3} \tag{1.90}$$

$$\omega_2^2 = \frac{V_{22}^E}{T_{22}^E} = \frac{18k - 6PL}{mL^3} \tag{1.91}$$

If we had been working only in the Q_i coordinates without the introduction of a scheme of diagonalization we could derive these frequencies alternatively from the characteristic equation

$$\begin{vmatrix} (5k - 2PL - \omega^2 \frac{2}{3}mL^3) & (-4k + PL - \omega^2 \frac{1}{6}mL^3) \\ (-4k + PL - \omega^2 \frac{1}{6}mL^3) & (5k - 2PL - \omega^2 \frac{2}{3}mL^3) \end{vmatrix} = 0 \tag{1.92}$$

which gives

$$(5k - 2PL - \omega^2 \tfrac{2}{3}mL^3) = \pm(-4k + PL - \omega^2 \tfrac{1}{6}mL^3) \tag{1.93}$$

and hence ω_1 and ω_2.

The normal mode linear vibration theory will be illustrated further by the eigenvalue analyses of vibrating struts and wires in the following chapter.

2

Vibration and Buckling of Beams and Struts

In this chapter we shall apply the general theory of Chapter 1 to the small-deflection linear vibration and buckling of continuous elastic beams and columns. These will be discretized by modal expansions of the harmonic and finite-element types, allowing the use of our earlier Lagrange equations.

The harmonic expansion yields the complete eigenvalue spectrum for the vibration and buckling of a strut with or without an elastic foundation, and the setting to zero of the bending stiffness allows us to retrieve the solution for a stretched string. An introductory polynomial analysis yields a two-degree-of-freedom vibration study of a cantilever.

The finite element method is introduced along the lines of our earlier treatment,[1] and as an example a four-mode buckling analysis of a pinned column is finally outlined.

2.1 BEAM FORMULATION

We shall make here an exact non-linear large-deflection energy formulation for the deformations of an elastic beam or column.[1,2] We shall then use the appropriate linearization for the eigenvalue analyses of the present chapter: the non-linear theory will be used in Chapter 4 to study the initial post-buckling behaviour of a column.

Consider the strut of Figure 2.1 of length L, simply supported to fix our ideas, and loaded by the axial force P which retains its magnitude and direction as the strut deflects. The strut is assumed to be axially rigid (inextensional) and the relevant bending stiffness is denoted by EI. Point A of the strut originally distance x from the left-hand support is displaced to B, and this displacement is resolved into an unspecified horizontal component and a vertical component w as shown. The centre-line being inextensional, the arc length SB is equal to x, and the deflected form of the strut is totally specified by the mathematical function $w(x)$, where x ranges from 0 to L. Notice that the graph of $w(x)$ does not have precisely the shape of the deflected beam because of the unspecified horizontal displacement.

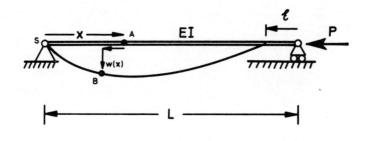

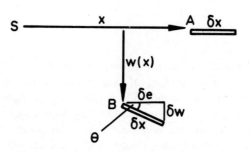

Figure 2.1 Notation for the large-deflection formulation of a strut

The curvature χ is by definition the rate of change of angle with arc length, so

$$\chi = \frac{d\theta}{dx} = \frac{d}{dx}\sin^{-1} w' = w''(1 - w'^2)^{-1/2} \tag{2.1}$$

where a prime denotes differentiation with respect to x. The strain energy stored in an element is

$$\delta U = \tfrac{1}{2}M\chi\delta x \tag{2.2}$$

where M is the bending moment given by

$$M = EI\chi \tag{2.3}$$

so

$$\delta U = \tfrac{1}{2}EI\chi^2\delta x \tag{2.4}$$

The total strain energy is thus

$$U = \tfrac{1}{2}EI\int_0^L \chi^2 dx$$

$$= \tfrac{1}{2}EI\int_0^L w''^2(1 - w'^2)^{-1}dx$$

$$= \tfrac{1}{2}EI\int_0^L (\underline{w''^2} + w''^2 w'^2 + w''^2 w'^4 + \cdots)dx \tag{2.5}$$

Here the familiar underlined term is all that need be retained for a small-deflection *linear* analysis.

From the figure we have

$$\delta e = (\delta x^2 - \delta w^2)^{1/2}$$
$$= \delta x(1 - w'^2)^{1/2} \tag{2.6}$$

so the end-shortening of the column is

$$\mathscr{E} = L - \int_0^L (1 - w'^2)^{1/2} dx$$

$$= \int_0^L (\tfrac{1}{2}w'^2 + \tfrac{1}{8}w'^4 + \tfrac{1}{16}w'^6 + \cdots) dx \tag{2.7}$$

Here the leading underlined term is again all that is needed for a linear buckling or vibration analysis.

Having made a precise static formulation we shall approximate some what in writing only the *linear form* of the kinetic energy. If the beam has the mass per unit length m, the kinetic energy of an element is approximately

$$\delta T = \tfrac{1}{2}m\delta x\dot{w}^2$$

where a dot denotes differentiation with respect to the time t, giving

$$T = \tfrac{1}{2}m \int_0^L \dot{w}^2 dx \tag{2.8}$$

2.2 MODAL EXPANSIONS

The linearized energy integrals necessary for small-amplitude vibration and buckling eigenvalue analyses of elastic beams are obtained from our non-linear formulation of Section 2.1 as follows.

The strain energy of bending is

$$U^B = \tfrac{1}{2}EI \int_0^L w''^2 dx \tag{2.9}$$

the potential energy of a dead axial compressive load P is

$$U^P = -P\mathscr{E} = -\tfrac{1}{2}P \int_0^L w'^2 dx \tag{2.10}$$

and the strain energy of a *simple* elastic foundation supporting the beam along its length can be written[1] as

$$U^F = \tfrac{1}{2}K \int_0^L w^2 dx \tag{2.11}$$

where K is the foundation stiffness. So our total potential energy V can be written as

$$V = U^B + U^F + U^P \tag{2.12}$$

and the kinetic energy, from equation (2.8), is finally

$$T = \tfrac{1}{2}m \int_0^L \dot{w}^2 dx \tag{2.13}$$

We propose to discretize the continuous elastic column in this section by employing a modal expansion for w,

$$w(x, t) = \sum_i q_i(t) M_i(x) \tag{2.14}$$

Here the amplitudes q_i are functions of time, and each mode-form M_i is assumed to satisfy the *geometric* boundary conditions of the problem. Each mode-form need not satisfy the *natural* or *dynamic* boundary conditions which will be satisfied automatically as far as possible by the subsequent energy procedures.

With a finite set of modes $M_i(x)$ we are thus employing the well-known Rayleigh–Ritz energy procedure, while if we employ a complete and necessarily infinite set of functions we have a classical exact analysis, assuming the convergence of certain infinite series.

Using the dummy-suffix summation convention we now have

$$\left.\begin{aligned} w = q_i M_i, \quad w' = q_i M_i' \\ w'' = q_i M_i'', \quad \dot{w} = \dot{q}_i M_i \end{aligned}\right\} \tag{2.15}$$

so we can write the products

$$\left.\begin{aligned} w^2 = q_i q_j M_i M_j, \quad w'^2 = q_i q_j M_i' M_j' \\ w''^2 = q_i q_j M_i'' M_j'', \quad \dot{w}^2 = \dot{q}_i \dot{q}_j M_i M_j \end{aligned}\right\} \tag{2.16}$$

The strain energy of bending is then

$$U^B = \tfrac{1}{2}EI q_i q_j \int_0^L M_i'' M_j'' dx \tag{2.17}$$

and comparing this with our standard linear form

$$U^B = \tfrac{1}{2}U_{ij}^B q_i q_j \tag{2.18}$$

we find the coefficients

$$U_{ij}^B = EI \int_0^L M_i'' M_j'' dx \tag{2.19}$$

Similarly, the potential of the load has coefficients

$$U_{ij}^P = -P \int_0^L M_i' M_j' dx \tag{2.20}$$

and the strain energy of the foundation has

$$U_{ij}^F = K \int_0^L M_i M_j dx \tag{2.21}$$

The potential energy coefficients are then

$$V_{ij} = U_{ij}^B + U_{ij}^F + U_{ij}^P \tag{2.22}$$

The kinetic energy is

$$T = \tfrac{1}{2}m\dot{q}_i\dot{q}_j \int_0^L M_i M_j \mathrm{d}x \tag{2.23}$$

and comparison with

$$T = \tfrac{1}{2}T_{ij}\dot{q}_i\dot{q}_j \tag{2.24}$$

gives us finally

$$T_{ij} = m \int_0^L M_i M_j \mathrm{d}x \tag{2.25}$$

The Lagrange equations for small vibrations (1.73) are

$$T_{ij}\ddot{q}_j + V_{ij}q_j = 0 \tag{2.26}$$

allowing a solution to our discretized beam problem.

2.3 COMPLETE HARMONIC ANALYSIS OF A COLUMN

As an example for a pin-ended beam we shall now take a complete Fourier expansion by writing

$$M_i = \sin\frac{i\pi x}{L} \tag{2.27}$$

for values of i from one to infinity. This greatly simplifies the analysis because the *orthogonality properties*

$$\left.\begin{array}{l} \displaystyle\int_0^L \sin\frac{i\pi x}{L}\sin\frac{j\pi x}{L}\mathrm{d}x = 0 \\[3mm] \displaystyle\int_0^L \cos\frac{i\pi x}{L}\cos\frac{j\pi x}{L}\mathrm{d}x = 0 \end{array}\right\} \quad \text{for } i \neq j \tag{2.28}$$

ensure that all the energies are diagonal. That is, the Fourier harmonics are, in fact, both the buckling and normal vibration modes of the simply-supported beam: the generalized coordinates q_i are already the principle coordinates u_i, and the equations of motions are already decoupled.

Then, since

$$\int_0^L \sin^2\frac{i\pi x}{L}\mathrm{d}x = \int_0^L \cos^2\frac{i\pi x}{L}\mathrm{d}x = \tfrac{1}{2}L \tag{2.29}$$

we have the diagonal energy coefficients

$$V_{ii} = \left\{ EI\left(\frac{i\pi}{L}\right)^4 + K - P\left(\frac{i\pi}{L}\right)^2 \right\}\tfrac{1}{2}L \tag{2.30}$$

$$T_{ii} = \{m\}\tfrac{1}{2}L \tag{2.31}$$

The equation for the circular frequency of the ith mode is therefore

$$\omega_i^2 = \frac{V_{ii}}{T_{ii}} = \left\{ EI\left(\frac{i\pi}{L}\right)^4 + K - P\left(\frac{i\pi}{L}\right)^2 \right\} \bigg/ m \tag{2.32}$$

This gives the normal frequencies of vibration of a simply-supported elastic beam of mass per unit length m and bending stiffness EI, resting on an elastic foundation of stiffness K and carrying an axial compressive load P.

If we set ω_i equal to zero (or equivalently set V_{ii} equal to zero) we have the critical buckling loads of this system as

$$P^i = \frac{EI\left(\dfrac{i\pi}{L}\right)^4 + K}{\left(\dfrac{i\pi}{L}\right)^2} \tag{2.33}$$

and in the absence of a foundation we can set $K = 0$ to obtain the critical loads of a pin-ended column

$$P^i = EI\left(\frac{i\pi}{L}\right)^2 \tag{2.34}$$

the lowest of which is

$$P^1 = \frac{\pi^2 EI}{L^2} \tag{2.35}$$

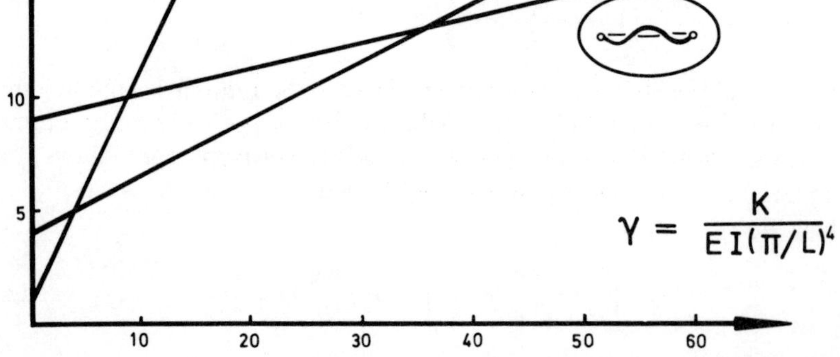

Figure 2.2 The variation of the first three critical buckling loads of a strut on an elastic foundation with the stiffness ratio γ

for the mode

$$M_1 = \sin\frac{\pi x}{L} \tag{2.36}$$

We see that *with* a foundation of stiffness K the first harmonic ($i = 1$) does not always correspond to the *lowest* buckling load, as illustrated in Figure 2.2.

If we put the foundation stiffness to zero and the bending stiffness to zero, $K = EI = 0$, and write $P = -S$, we have the formula for the normal frequencies of a string stretched by a tension S as

$$\omega_i^2 = S\left(\frac{i\pi}{L}\right)^2 \Big/ m \tag{2.37}$$

2.4 VIBRATION OF A CANTILEVER IN TWO MODES

Just as we can make one-degree-of-freedom Rayleigh–Ritz beam analyses using simple assumed forms, so we can perform multi-mode studies, and we here make a two-degree-of-freedom study of the vibrations of a cantilevered beam as an introduction to the finite element method.

Treating the cantilever of Figure 2.3 essentially as one element, we give its tip a free deflection Q_1 and a free slope Q_2 as shown. A polynomial assumed form for the deflected shape which satisfies the clamped boundary is

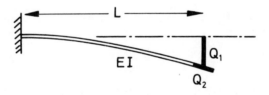

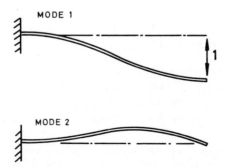

Figure 2.3 A two-mode vibration analysis of a cantilevered beam, showing the two mode forms

$$w(x) = Q_1 \left\{ 3\left(\frac{x}{L}\right)^2 - 2\left(\frac{x}{L}\right)^3 \right\} + Q_2 L \left\{ -\left(\frac{x}{L}\right)^2 + \left(\frac{x}{L}\right)^3 \right\} \qquad (2.38)$$

where the braces are our mode forms M_1 and M_2. These have the form shown in the figure, mode one being found by setting $Q_1 = 1$ and $Q_2 = 0$, etc.

Substituting into the energy integrals, the potential energy and kinetic energy coefficients are readily determined as

$$V_{ij} = \frac{EI}{L^3} \begin{bmatrix} 12 & -6L \\ -6L & 4L^2 \end{bmatrix} \qquad (2.39)$$

$$T_{ij} = \frac{mL}{210} \begin{bmatrix} 78 & -11L \\ -11L & 2L^2 \end{bmatrix} \qquad (2.40)$$

To determine the frequencies of vibration of this unloaded cantilever we can now follow the non-diagonalized scheme of analysis and write the characteristic determinantal equation

$$|V_{ij} - \omega^2 T_{ij}| = 0 \qquad (2.41)$$

This is expanded to give

$$D = (12 - 0.371W)(4 - 0.00952W) - (0.0524W - 6)^2 = 0 \qquad (2.42)$$

where

$$W = \frac{mL^4}{EI} \omega^2 \qquad (2.43)$$

and plotting D against W to find its zeros, we obtain for the fundamental frequency of vibration

$$\omega^2 = \frac{EI}{mL^4} 12.5 \qquad (2.44)$$

This compares with the exact fundamental frequency which has the same form with a numerical factor of 12.36.

Such a two-degree-of-freedom analysis gives also an estimate of the second frequency.

2.5 THE FINITE ELEMENT METHOD

We shall view the finite element method,[1] as a way of patching together a form for the deflection function $w(x)$ of a beam as shown in Figure 2.4. This is done strictly to a mathematical plot of w against x, which does not represent precisely the physical shape of the deformed beam because of the unspecified horizontal component of deflection.

Displacement and slope are specified as free variables, or generalized coordinates, at the internal nodes, and in each region of the beam we fit a third-order polynomial of the form

$$w(x) = a + bx + cx^2 + dx^3 \qquad (2.45)$$

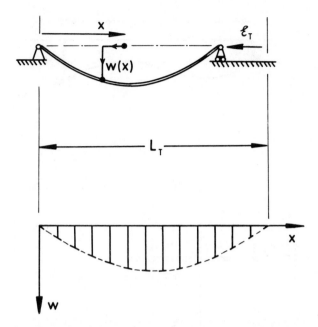

Figure 2.4 The finite element representation of the deflection function $w(x)$ for a pin-ended beam or column

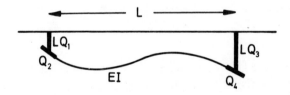

$$\frac{L}{EI}\left[U_{ij}\right] = \begin{bmatrix} 12 & 6 & -12 & 6 \\ 6 & 4 & -6 & 2 \\ -12 & -6 & 12 & -6 \\ 6 & 2 & -6 & 4 \end{bmatrix}$$

Figure 2.5 The stiffness matrix for an element of length L

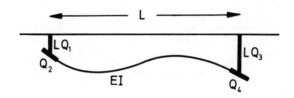

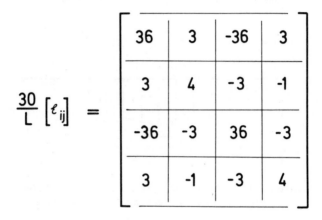

Figure 2.6 The end-shortening matrix for an element of length L

Taking now $0 < x < L$ as the coordinate of an element of length L we can find $a, b, c,$ and d in terms of the end-variables, $Q_1, Q_2, Q_3,$ and Q_4 of Figure 2.5. They can be written as

$$a = Q_1 L, \, b = Q_2$$
$$c = \{-3Q_1 - 2Q_2 + 3Q_3 - Q_4\}/L \qquad (2.46)$$
$$d = \{2Q_1 + Q_2 - 2Q_3 + Q_4\}/L^2$$

Integrating over this element, we then obtain, after some tedious algebra, the strain energy matrix of Figure 2.5 and the end-shortening matrix of Figure 2.6.

For two consecutive elements these matrices can be combined as shown in Figure 2.7, with an overlap due to the sharing of coordinates Q_3 and Q_4.

We note that the same fundamental principles can be extended to higher-order terms for use in post-buckling study, as described and fully illustrated in our earlier text.[1] Three-, four-, and six-dimensional matrices, representing cubic, quartic, and sixth-order terms of energy, are employed in various strut, beam, and frame problems.

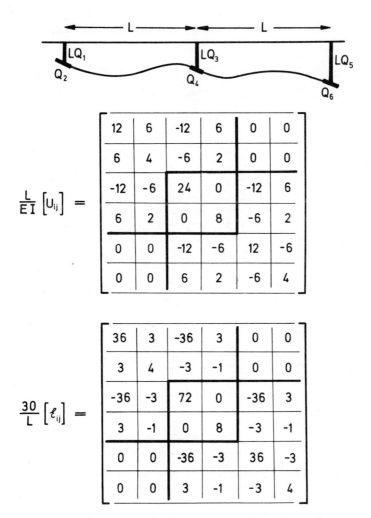

Figure 2.7 The assembly of the stiffness and end-shortening matrices for two consecutive elements of a beam

2.6 BUCKLING OF A STRUT IN FOUR MODES

We give finally an example of the use of this finite-element procedure for the *linear buckling* of the pin-ended strut of Figure 2.8. With two elements each of length L we have here four degrees of freedom represented by Q_A, Q_B, Q_C, and Q_D, there being no displacement allowed at the pinned supports. The V_{ij} matrix is obtained from

$$V_{ij} = U_{ij} - P\mathscr{E}_{ij} \qquad (2.47)$$

by deleting the rows and columns of Q_1 and Q_5 in Figure 2.7 with the notation

38

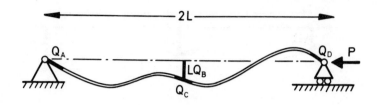

$$\frac{L}{EI}\left[V_{ij}\right] = \begin{bmatrix} 4-4\Lambda & -6+3\Lambda & 2+\Lambda & \text{zero} \\ -6+3\Lambda & 24-72\Lambda & \text{zero} & 6-3\Lambda \\ 2+\Lambda & \text{zero} & 8-8\Lambda & 2+\Lambda \\ \text{zero} & 6-3\Lambda & 2+\Lambda & 4-4\Lambda \end{bmatrix}$$

$$\boxed{\Lambda = \frac{PL^2}{30EI}} \qquad \underline{\Lambda^1 = 0.08287} \qquad \Lambda^2 = \frac{2}{5}$$

$$\Lambda^3 = 1.073 \qquad \Lambda^4 = 2$$

$$\underline{\Lambda^1_{\text{EXACT}} = \frac{\pi^2}{120} = 0.08225}$$

Figure 2.8 A two-element analysis of a pin-ended strut buckling in four degrees of freedom

$$\Lambda = \frac{PL^2}{30EI} \tag{2.48}$$

The equilibrium equations for this strut are

$$\frac{\partial}{\partial Q_s}(\tfrac{1}{2}V_{ij}Q_iQ_j) = V_{sj}Q_j = 0 \tag{2.49}$$

and these *decompose* into a physically symmetric solution with $Q_C = 0$ and $Q_A = -Q_D$ and a skew-symmetric solution with $Q_B = 0$ and $Q_A = Q_D$. The vanishing of the determinant of the symmetric solution yields

$$45\Lambda^2 - 52\Lambda + 4 = 0 \tag{2.50}$$

which gives the eigenvalues for Λ. In terms of the total length $L_T = 2L$ these are

$$\frac{PL_T^2}{EI} = 9.944 \quad (\pi^2 = 9.87) \tag{2.51}$$

$$\text{and } 128.6 \quad (9\pi^2 = 88.8) \tag{2.52}$$

with the exact answers shown in brackets. We see that this rudimentary two-element analysis has supplied a fairly good estimate of the first harmonic buckling load but a poor estimate of the third harmonic buckling load.

The vanishing of the determinant of the skew-symmetric solution yields

$$(4 - 4\Lambda)^2 - (2 + \Lambda)^2 = 0 \tag{2.53}$$

giving

$$\frac{PL_T^2}{EI} = 48 \quad (4\pi^2 = 39.5) \tag{2.54}$$

$$\text{and } 240 \quad (16\pi^2 = 158) \tag{2.55}$$

which are again compared with the exact figure in brackets.

We see that we always get as many eigenvalues as degrees of freedom, but that the accuracy of the solutions decreases rapidly for the higher modes. High degrees of accuracy can, of course, be achieved by the use of a large number of elements, as demonstrated clearly in our earlier book.[1]

3

Loads and Imperfections

In the study of conservative mechanical systems in Chapter 1 we looked at the behaviour of a *given system* governed by $V(Q_i)$, and wrote down two fundamental axioms relating its equilibrium and stability to the form of this potential energy function. The first states that a stationary value of the total potential energy is necessary and sufficient for equilibrium. The second states that a complete relative minimum of the energy is both necessary and sufficient for stability of an equilibrium state.

We remember that the proof of the second axiom involves an appeal to the presence of a little positive definite viscous damping which will always be assumed to be present in our future discussions. Indeed, a trace of damping is essential for good modelling of the real world, and its inclusion brings *gyroscopic systems* within the embrace of our two axioms.[2]

In the present chapter we introduce a wide body of theory that draws on these two axioms to classify the response of a *family of systems* generated by the insertion of control parameters Λ^j, usually loads and imperfections, into the energy function, so that we have $V = V(Q_i, \Lambda^j)$. Before proceeding with this, we briefly refer to a number of fields in which our two axioms hold, going beyond the general mechanical system of our first chapter.

3.1 RELATED FIELDS OF APPLICABILITY

The two axioms, and the theory that follows, relate to any system that seeks out local minima of a function $V(Q_i)$ whether or not this function has any relationship at all to energy in general, or potential energy in particular.[2]

Historically, the axioms and the resulting bifurcation theory were used extensively in early investigations into the stability of rotating liquid masses which were used to model primitive planets.[2,14] This is an example of their use for gyroscopic systems involving coriolis forces, where they are valid in the presence of some damping.

The axioms hold in many situations in thermodynamics, where the appropriate potential function might be the Gibb's free energy, the Helmholtz free energy, the enthalpy, or the negative of the entropy.[15] The theory can thus be used in discussing the stability of stars and stellar clusters.[2,16,17] In astrophysics they apply, moreover, to a unique mechanics problem in the general theory of relativity, where the appropriate potential is the mass energy of a massive cold star.[2,18]

The general bifurcation theory of our first book[1] has been used in crystallography, where a Newtonian approximation to the quantum mechanical problem is valid for the evaluation of the bulk mechanical properties of a solid.[19] Here the bifurcational instability of an atomic lattice is studied using the Lennard–Jones interatomic potentials.[2,20,21] New related work of Hill[22] in finite elasticity is relevant to metal crystals and rubber-like polymers.

Other well-defined instabilities governed by a potential function arise in meniscus studies, including the collapse of water drops and air bubbles.[23] A fascinating electrohydrostatic problem is the loss of symmetry and stability of a charged water drop, relevant to the onset of thunderstorms.[24]

We mention finally two applications of catastrophe theory in physics, where the governing function does not have the nature of energy.

Closely analogous to Hamilton's principle in mechanics is Fermat's so-called principle of 'least-time' in ray optics. This states that light seeks to minimize the time taken to travel between two distinct points, and thus embraces both the laws of reflection and refraction, as well as the more fundamental fact that it travels in a straight line in a homogeneous medium. But this is not to suggest that a ray literally economizes its journey time; indeed, we should more correctly refer to the principle of 'stationary-time', since maxima and saddle-points give equally valid paths. To understand the principles fully we must see light not as particles but as waves. Only along paths of stationary time do adjacent waves reinforce one another. For a very readable account of this, along with a more general discussion of variational principles in quantum mechanical terms, we refer to the excellent book of Poston and Stewart.[5] Such considerations have led to one of the most significant applications of catastrophe theory to date, in Berry's and Nye's detailed studies of light caustics.[25–30]

Second, Berry and Mackley[31] have used the classified structurally-stable singularities to interpret phenomena in the steady two-dimensional flow of a fluid. Here there is no potential function as such, but an analogous *stream function* which has as its contours the stream-lines of the flow pattern. Maxima, minima, and saddles relate to stagnation points in the flow where the velocity is zero, and each will clearly have its own typical form; but non-stationary points are now likewise of crucial importance in determining the overall flow pattern. In this context, Berry and Mackley use the elliptic umbilic singularity to unfold an unstable fluid flow with the generation of a circulating vortex. Benjamin's fine experimental study of generic bifurcations in fluid flow should also be mentioned here.[32]

Analogous to the stream function is the stress function of elasticity, and singularities in vector fields relevant to the interpretation of photoelastic patterns have been examined in catastrophe theory terms by Nye and co-workers.[33,34]

Before returning to the main development of this chapter we emphasize once again that non-conservative problems arising, for example, from fluid loading are quite outside the scope of the present study based on a potential function. The associated dynamic flutter instabilities, which are not governed by simple potential energy considerations, are discussed extensively in reference 2.

Comprehensive reviews of applications of catastrophe theory in the physical sciences are due to Thompson and Hunt[35] and Stewart.[36] Along with the books already mentioned,[1-5] we would also draw attention to those of Saunders[37] and Gilmore.[38]

3.2 ELIMINATION OF PASSIVE COORDINATES

We return to the main development of this treatise, and its general conservative mechanical system whose equilibrium and stability are governed by $V(Q_i)$, with i ranging from one to n.

Before considering the variation of loads in detail we suppose here that by some evolutionary process the system has arrived at an m-fold critical equilibrium state, where m stability coefficients vanish simultaneously; we assume that no critical states of order greater than m can be found in the region of interest, usually a subset of all attainable states, local to the m-fold state. In a buckling situation this will mean that we have *simultaneous buckling* in m buckling modes at a critical state of loading, and that there are no other buckling loads close by. Of course, we are including in this discussion distinct buckling in a single mode, described by $m = 1$, as in the buckling of an Euler column.

Now engineers are familiar with the idea that at a point of incipient instability it is the buckling modes themselves that dominate the behaviour of the structure: other non-critical modes (such as the second, third, and higher harmonics of a strut) will play, if anything, a rather secondary passive role. In this spirit a structural engineer might make a Rayleigh–Ritz analysis using *only* the m coincident buckling modes to analyse the buckling and initial post-buckling behaviour.

Now while it is often not correct to just use such a Rayleigh–Ritz procedure (see for example, the effect of passive modes on the initial post-buckling path curvature in the plate buckling model of the following chapter) it is true that in such a situation the passive coordinates can be eliminated more skilfully from our consideration once and for all. We have shown this in detail in our first book,[1] following the lead of Koiter's thesis.[12] We must emphasize that to achieve this elimination of passive coordinates we do *not just strike them out*, as is effectively done in our discussed Rayleigh–Ritz procedure. They are rather made functions of the active coordinates, so that they change in a slave-like manner implicitly throughout the analysis.

Mathematically, we can extract a local m-dimensional problem from the global n-dimensional one, to study the instability in isolation. This can have considerable analytical and conceptual advantages, since in most practical circumstances (with a few notorious exceptions) m will be small even though n might be very large or strictly infinite; at a simple distinct eigenvalue we have, of course, $m = 1$.

We refer to this process as the *elimination of passive coordinates*, and note that it has an impeccable mathematical pedigree based on the implicit function theorem.

In Chapter 7, under a bifurcational formalism, we shall present a perturbation scheme which can perform this task in a systematic manner. We segregate the n generalized coordinates into two groups, m of which take a major part in the instability and are termed *active coordinates*, the remaining $n-m$ being the passive coordinates. As we have emphasized, this does not mean that the passive coordinates take no part (indeed, *quantitatively* they may be just as significant as the actives, and influence our numerical answers through the 'contamination' effects discussed later), but once they are given by the scheme as functions of the active coordinates they can be effectively dropped from direct consideration in the analysis; this is not the same thing as just crossing them out, as is sometimes done incorrectly in buckling studies.

Thinking physically, the coordinates must be so segregated that the placing of constraints on the active coordinates would be sufficient to inhibit the instability. The mathematical condition for this is that the sub-determinant of the passive coordinates should be non-zero, and it is a property of the full singular determinant of rank $n-m$ that a valid segregation can always be found for a given set of generalized coordinates: the segregation, however, may not be unique, but we simply assume that one valid segregation has been chosen.

It must be stressed that the elimination of passive coordinates is usually only performed *locally*, and in describing the complete *global* behaviour of a system we might well need all n generalized coordinates to vary independently. This raises the interesting question of just what we mean by 'local', and that may depend on a number of other factors, such as relative complexities of analysis. We shall consider later two bifurcation points nearly but not completely coincident. Each could be analysed separately, or we could treat the case as a perturbed version of complete coincidence; the latter is more complex, but will have a greater range of validity.

3.3 LOSS OF STABILITY UNDER LOAD

Up to this point we have considered all systems to be frozen in evolutionary time, but we now allow the potential surface to distort slowly, and study the simple *critical* states that can thus arise. We do this by introducing controlled variations into the system, and write the potential function

$$V = V(Q_i, \Lambda^j) \tag{3.1}$$

where Q_i is the set of n generalized coordinates as before and Λ^j is a set of h control parameters. These can be genuinely controlled, as with a load on a model structure, or they may be the magnitudes of imperfections or other perturbations in the family of systems. This latter type can, of course, arise in a variety of ways, and it is here that catastrophe theory has added to earlier bifurcation theories, since it tells us just which are of major significance in a given situation.

Some distinct critical points

We outline in this section some simple modes of behaviour at distinct critical points with a single active coordinate Q ($m = 1$, it being understood that we have suitably eliminated any other passive coordinates) and with $h = 1$ for a single control parameter Λ, which can be thought of as a load.

The most typical situation is included in Figure 3.1, which shows schematically how a potential energy curve with two minima separated by a maximum can be smoothly transformed into a curve with just a single minimum. The essential local transformation at C involves just the coalescence and annihilation of a maximum and a minimum at a horizontal point of inflexion.

The variation of the potential curve with Λ must be infinitely slow in comparison with the dynamic response of the system, so that we can always think of the potential curve as fixed while the system settles to a stable equilibrium state. Essentially we consider the motion of the system at different but constant values of Λ, just as our basic axioms relate to tests of V at constant Λ. We cannot therefore deal with any of the considerable difficulties which arise when the evolutionary process (loading) and dynamic effects are seen on the same time-scale, even though we know, for example, that a structure will have a very sluggish response near a critical equilibrium state because the damping inevitably becomes super-critical as a stability coefficient tends to zero (reference 2, Figure 5). Such considerations are not within the scope of this book.

Because our stability axiom only specifies a *local* minimum for stability, we are

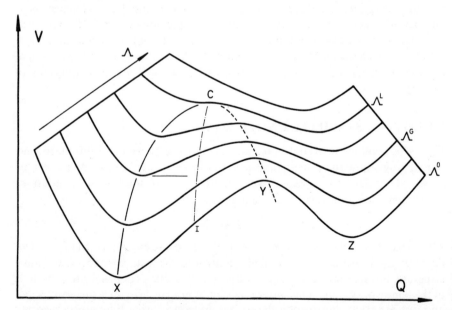

Figure 3.1 A parametrized total potential energy function $V(Q, \Lambda)$, showing how under a single control parameter Λ a horizontal point of inflection can be generated in a typical manner to give a fold or limit point

concerned throughout the book with what is called the *delay convention*. The alternative *Maxwell* convention, applicable in certain other disciplines, demands a *global* minimum for stability with the result that meta-stable states (according to the delay rule) are declared unstable. Thus as Λ is increased in Figure 3.1 the loss of Maxwell stability would occur at $\Lambda = \Lambda^G$, while for our purposes the loss of stability is delayed until $\Lambda = \Lambda^L$. Clearly, a parallel treatment exists, but this is all we shall say about the Maxwell convention.

As Λ of Figure 3.1 increases from Λ^0 we see that in the space of Λ and Q (hinted at in this schematic diagram) we have a rising stable equilibrium path, shown as a solid line. This reaches a maximum at C, where it becomes unstable (denoted by a broken line) and falls. Thus our mechanical system following this path under prescribed increasing load becomes unstable on reaching C and jumps dynamically to the remote single minimum.

This limit point, or fold catastrophe, about C arises right across the spectrum of the sciences, and is *the* typical mode of failure for a system under a single control, as we shall see later. Thus a real (and therefore necessarily imperfect) structure under load will always fail at such a limit point, but it is often the relationship that this critical state has with some underlying more complex singularity (such as a point of bifurcation of an idealized perfect structure) that is of crucial significance. This is a theme to which we constantly return.

We discuss next three common distinct points of bifurcation, where two equilibrium paths intersect: these are shown in Figure 3.2. Here we are plotting for each the load Λ against an incremental coordinate q, measured always from the rising initially-stable (primary or fundamental) equilibrium path: this has the effect of making the fundamental path coincident with the load axis. Stable paths are shown as solid lines, while unstable paths are denoted by broken lines. The variations of the total potential energy with q at various fixed Λ levels are shown, and we can think of the system as a ball rolling on these curves: in accordance

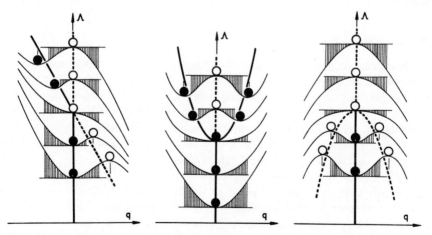

Figure 3.2 Total potential energy transformations in the asymmetric, stable-symmetric, and unstable-symmetric points of bifurcation

with this, we show a black ball at each stable state and a white ball at each unstable state.

In the first *asymmetric* point of bifurcation the primary path loses its stability as it intersects an inclined secondary (post-buckling) equilibrium path: these two paths exhibit the exchange of stabilities discussed by Poincaré.[39] Here we see that a minimum and a maximum of V coalesce, as in the limit point, but emerge again as maxima and minima above the critical point. We shall see later that this asymmetric (or transcritical) branching point is, in fact, a rather specialized view of the fold catastrophe.

Next consider the *stable-symmetric* point of bifurcation of the middle diagram. Here the primary path once again becomes unstable as it intersects a secondary path, but the latter has a zero slope and positive curvature at the bifurcation point, so it curves upwards and is stable. The localized $V(q)$ plots indicate that we have a minimum of V in the large, but above the critical point a local maximum has emerged, bringing with it a local minimum on either side. This stable-symmetric or super-critical bifurcation (sometimes called the pitchfork) corresponds to the cusp catastrophe.

Finally consider the opposite situation in the last picture of Figure 3.2. Again we have a primary path becoming unstable at the critical point, but here the secondary post-buckling path has a zero slope with negative curvature, and is unstable. The $V(q)$ plots indicate a maximum of V in the far field, but below the critical point we have a localized minimum with its associated maxima. This *unstable-symmetric* or sub-critical point of bifurcation is really just an inversion of the previous case, and corresponds to the dual cusp.

These four patterns of behaviour, the limit point and the three distinct bifurcations, can be found in a number of texts, spanning many disciplines, particularly in the physical sciences. They are clearly the most likely forms of instability to arise, and a very thorough treatment of them is given in our original book.[1,40]

Two basic theorems

We have seen the most simple instability phenomena that can arise, and it seems appropriate before introducing imperfections to have a look at two general results that are hinted at by these four forms. Are there any specific conclusions we can draw that cover any such paths, including, perhaps, some very pathological critical states? We apply ourselves to this question by restricting attention, as before, to a general conservative system with a single control parameter, governed by the single-valued potential $V(Q_i, \Lambda)$.

The first point that can easily be proved[1] is that there is a unique equilibrium path through any non-critical point at which $|V_{ij}| \neq 0$. With the question of stability very much in mind, Thompson in 1970 formulated two basic theorems,[41] related to the passage of equilibrium paths through critical equilibrium states. Until recently, proofs were only available for systems with a single active coordinate,[1,41] although we remained convinced of their truth in general.

However, in 1976 Kuiper and Chillingworth proved the two theorems for the m-fold critical state, using techniques of modern topology:[42] this is a good example of our fruitful interaction with catastrophe theory over the period from 1975. They can thus be regarded as proven for our general system, and we can state them as follows:[1]

Theorem 1 An initially-stable (primary) equilibrium path rising monotonically with a single control parameter Λ cannot become unstable without intersecting a further distinct (secondary) equilibrium path.

Theorem 2 An initially-stable equilibrium path rising with a single control parameter Λ cannot approach an unstable equilibrium state from which the system would exhibit a finite dynamic snap without the approach of an equilibrium path (which may or may not be an extension of the original path) at values of the loading parameter less than that of the unstable state.

The circumstances ruled out by these two theorems are illustrated in Figures 31 and 32 of reference 1, where they are discussed at length. They are phrased in general global terms, and are thus essentially stronger statements than the more common linear eigenvalue theorems based on the existence of an 'adjacent' position of equilibrium.

3.4 IMPERFECTIONS AND PERTURBED BIFURCATIONS

It has been a general conceptual viewpoint in many of the physical sciences, including structural engineering, to suppose that instabilities are primarily generated by the variation of a single control parameter, Λ. Secondary control parameters, often though not necessarily imperfections, are then introduced once the basic behaviour under Λ has been analysed.

A single basic control parameter leads naturally to the consideration of *equilibrium paths* in the $\Lambda-Q_i$ space, and introduction of secondary 'imperfection' parameters, imagined to be held constant at small values, is then seen to perturb the original set of equilibrium paths. We will refer to this as a *bifurcational formalism* or viewpoint. Now, of course, it was long realized that all control parameters *could* be viewed as equals, leading to a discussion of equilibrium surfaces in a full Λ^j-Q_i space, as in the work of Sewell, Huseyin, and others.[43,44] But it was not until the flowering of catastrophe theory that the drawing of equilibrium surfaces became universally fashionable. For this reason we shall call the unified view, in which all control parameters are regarded as equals, as the *catastrophe theory* viewpoint.

The distinction we are making here is not just one of how we choose to plot our results, but runs deeper. In catastrophe theory, since the parameters are regarded as equals, transformation of axes in the Λ^j control space are allowable even though they may mix the load Λ with the imperfections. This means that singularities that look identical under the catastrophe theory view may look quite different in bifurcationally orientated situations. We drew attention to this in 1975,[45] where we showed that a bifurcational viewpoint led to a finer classification than that of Thom's catastrophe theory. However, the essential

48

ingredient of catastrophe theory is not the unification of the control parameters but the focusing on singularities that are in a topological sense 'structurally stable'. This ingredient has now been used to produce a classification of structurally stable bifurcational instabilities by Golubitsky and Schaeffer in an important paper.[6]

In this section we shall now look in a simple manner at the effect of imperfections in a bifurcational setting. We shall not attempt at this stage to discuss in depth the underlying reason for the importance of imperfection parameters. These arise in a great variety of ways, and are sometimes, but not always, of crucial significance. It suffices to say that the great success, after their introduction by Koiter in 1945,[12] was to explain serious discrepancies between theoretical predictions and experimental results, coupled with heavy scatter of the latter, that had arisen in shell buckling studies. In the following sections we shall see that *imperfection-sensitivity* is closely related to the topological structural instability of certain bifurcational forms, and the wider implications in relation to symmetry, optimization, and design of real systems[46] are discussed in Chapter 8.

We consider, then, a system described by the potential function

$$V = V(Q, \Lambda, \varepsilon) \tag{3.2}$$

subject to our two axioms at constant Λ and ε, where Q is a single active coordinate and Λ and ε are control parameters. With our bifurcational viewpoint we shall speak of Λ as a loading parameter and ε as an imperfection parameter describing some perturbation of the system: ε could, for example, be the amplitude of an initial deformation, or the magnitude of an initial pre-stress. Thus if $\varepsilon = 0$, V describes the behaviour of some *perfect system*, and with the variation of ε we can generate an associated family of imperfect systems.

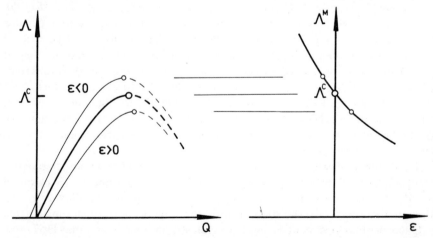

Figure 3.3 The effect of initial imperfections (perturbations) on the equilibrium path of a fold or limit point. The right-hand diagram shows the mild, non-singular, imperfection-sensitivity

Let us take first the example of the limit point discussed earlier. The equilibrium paths in Λ–Q space of the family of systems is shown in Figure 3.3. We can see that the presence of imperfections here makes very little difference to the response of the system: both the perfect system, corresponding to a heavy line, and the imperfect systems, corresponding to light lines, lose their stability at limit points, the local topology remaining unchanged by ε. On the right-hand side we show the *imperfection-sensitivity* or the variation of the *critical load* with *imperfection*, giving us the failure locus in control space; this is locally of a non-singular linear form.

But the same is not true for the asymmetric point of bifurcation, shown in Figure 3.4. Here the picture can be very much altered by the presence of an imperfection, and in two different ways. For a positive value of ε, say, the intersecting paths of the perfect system become two disconnected paths exhibiting limit points, while for a negative value of ε they are again disconnected but exhibit no critical points, as shown. The localized V–Q plots are not given explicitly on the figure, but with a little careful thought can be visualized as perturbed versions of the plots shown in Figure 3.2.

The imperfection-sensitivity plot shown on the right-hand side is now of much more significance. The curve is locally parabolic, exhibiting an infinite slope at $\varepsilon = 0$, so the load-carrying capacity of the structure is noticeably eroded by a very small positive imperfection; on the other hand a negative imperfection allows the system to escape the instability entirely. The imperfection parameter is thus considerably more important to this system than one which displays just a limit

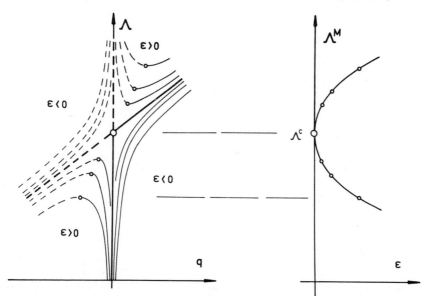

Figure 3.4 The effect of initial imperfections on the asymmetric point of bifurcation. The left-hand diagram shows the equilibrium paths and the right-hand diagram the severe imperfection-sensitivity

50

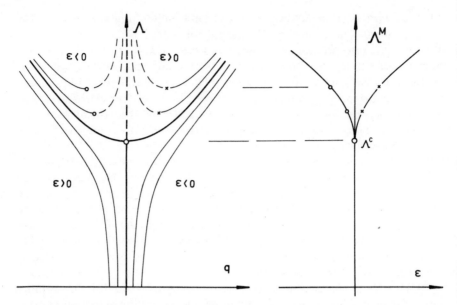

Figure 3.5 The effect of initial imperfections on the stable-symmetric point of bifurcation, showing the equilibrium paths and the stability boundary relating to the complementary equilibrium paths

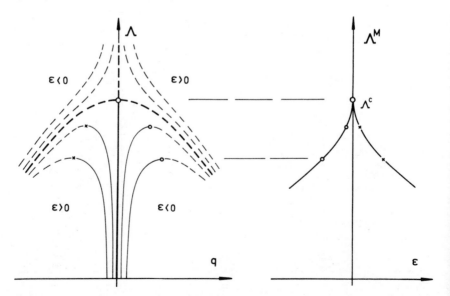

Figure 3.6 The effect of initial imperfections on the unstable-symmetric point of bifurcation, showing the equilibrium paths, and the severe imperfection-sensitivity with its two-thirds power law cusp

point, and we have here a hint of the concept of structural stability, discussed in the next section.

Likewise for the stable-symmetric point of bifurcation, the topological form of intersecting paths of the perfect system is destroyed by a symmetry-breaking imperfection ε, as shown in the left-hand plot of Figure 3.5. For a positive or negative value of ε, we have a continuous stable equilibrium path, always rising, but which brings with it an associated complementary path in the opposite quadrant.

Now we see that for all imperfect systems the *natural* loading path, which emerges from close to the origin $\Lambda = q = 0$ of Figure 3.5, is always rising and stable, so a stable-symmetric point of bifurcation exhibits no imperfection-sensitivity in the normally accepted sense. However, limit points do arise in the *complementary* equilibrium paths, and we can project these onto the control space $\Lambda-\varepsilon$ to give the stability boundary shown in the right-hand diagram. This is locally a two-thirds power law cusp, and we see the reason for naming this phenomenon the cusp catastrophe.

Finally, consider the unstable-symmetric point of bifurcation, shown in Figure 3.6. Here we have the opposite situation, with stable paths replacing unstable paths (and vice versa) and the direction of the Λ-axis reversed. The limit points now appear in the natural loading paths of imperfect systems, and we have the two-thirds power law cusp shown on the right-hand side; this again has an infinite slope at $\varepsilon = 0$, which implies considerable erosion of load-carrying capacity, this time for positive or negative ε.

Computed curves for the stable and unstable symmetric bifurcations are

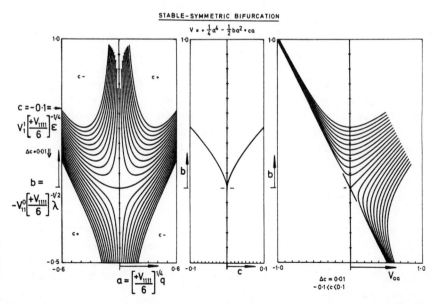

Figure 3.7 Computed universal graphs for the stable-symmetric point of bifurcation, corresponding to the displayed equation

52

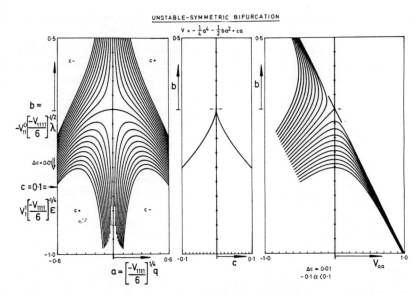

$$V = -\tfrac{1}{4}a^4 - \tfrac{1}{2}ba^2 + ca$$

Figure 3.8 Computed universal graphs for the unstable-symmetric point of bifurcation, corresponding to the displayed equation

shown in Figures 3.7 and 3.8. These are drawn for the total potential energy functions $V(a, b, c)$ shown, where a is a suitably scaled coordinate (corresponding to q), b is a suitably scaled load (corresponding to $\lambda = \Lambda - \Lambda^C$), and c is a suitably scaled imperfection (corresponding to ε). The nature of this scaling, indicated with reference to energy derivatives defined elsewhere, is such that all possible bifurcations of this type can be represented on these single *universal* local diagrams. Notice, in particular, that the curves of the stable and unstable symmetric bifurcations are numerically identical. For each bifurcation the left-hand diagram shows the load-deflection curves and the centre diagram the stability boundary boundary (imperfection-sensitivity curve) in control space. The right hand diagram shows the variation of the local stability coefficient $V_{aa}(\partial^2 V/\partial a^2$ evaluated at the relevant equilibrium state) with b; this coefficient is relevant to the study of the local frequency of linear vibrations since from equation (1.78) the circular frequency ω is given by

$$\omega^2 = \frac{V_{aa}}{T_{aa}} \tag{3.3}$$

where the kinetic energy coefficient T_{aa} can usually be presumed to be a *constant* in the vicinity of a bifurcation point. Watching the dropping towards zero of the lowest natural frequency of a structure is a useful non-destructive experimental technique for predicting the approach towards a critical load, without actually buckling the specimen.

These diagrams complete the bifurcational, or equilibrium path, view of the distinct critical points, and have provided the basis for a considerable volume of analytical study.[1,12,47-52] They were first confirmed experimentally by Roorda[53-6,] who introduced, in arch and frame buckling models, the variable load off-set as a controlled imperfection parameter. We note in passing that the patterns of equilibrium paths shown here can, of course, be taken as contours of a general equilibrium surface in three dimensions.

3.5 ELIMINATION OF PASSIVE CONTROLS

Having introduced the idea of small perturbations as controls, we are now faced with a description $V(Q_i, \Lambda^j)$ of the system, with the number h of control parameters possibly very large; we might say that we can perturb the system in a great many, if not an infinite, number of ways. We can usually reduce the number of generalized coordinates via the elimination of the passives, and it would be convenient to find a means of eliminating some of the controls. Quantitative elimination is clearly out of the question, for then we would have to specify *all* possible perturbations, but a qualitative elimination can often be achieved using the topological concept of *structural stability*; this is the key to the contribution of catastrophe theory.

Before we proceed we must mention a potential source of confusion in the terminology. We are mainly concerned in this book with the stability of elastic systems, and this is sometimes referred to in the engineering literature as 'structural stability', although here neither 'structure' nor 'stability' takes the same meaning as in the topological context. Since both usages have become commonplace in their respective disciplines we see no neat solution to this problem. However, we shall reserve the distinctive phrase 'structural stability' for the topological context, otherwise paraphrasing it as, for example, 'the stability of elastic structures' when the other meaning is intended. This is by no means ideal, but it must suffice.

To continue, structural stability suggests that for a topological structure to exist in the real natural world the entire phenomenon must possess global stability in the face of sufficiently small perturbations. Thus, for all three bifurcation points of the previous section the form of two intersecting paths of the perfect system is structurally unstable, since it is destroyed by an arbitrarily small imperfection, as we have described. In contrast, with a single Λ-parameter the limit point is a structurally stable phenomenon—imperfections only generate further limit points—and the imperfection-sensitivity is mild and non-singular.

We can gain more insight into the concepts involved by looking at the limit point in greater depth.[57] Consider a one-dimensional potential surface like that shown in Figure 3.9. Here we would certainly not be surprised to find non-critical equilibrium states (Figure 3.9 has three, X, Y, and Z), but critical equilibrium states would be quite a different matter; if, for example, we model such a surface by an appropriate but random scrawl on a blackboard or piece of paper, maxima and minima can arise in quite typical fashion, but a horizontal point of inflexion would not normally be found. This is reflected in the fact that to describe such a point we need two equations ($dV/dQ = 0$ for equilibrium and $d^2V/dQ^2 = 0$ for

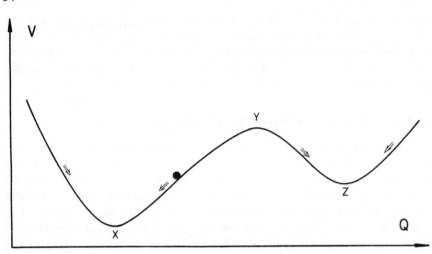

Figure 3.9 A typical un-parametrized $V(Q)$ potential energy curve with two stable equuilibrium states and one unstable equilibrium state

criticality), but we have only one unknown, Q; introducing a second coordinate is no help, since it would simply bring with it a second equilibrium equation. Thus, without parametric variation, the horizontal point of inflexion can be regarded as, in a definable sense, non-typical.

It is also structurally unstable, since with, say, a small tilt of the surface we can generate either a maximum and a minimum, or no equilibrium state. For a discussion in a mathematical context of the related concepts of genericity (typicality) and structural stability, including the differences identified by Smale in 1966,[58] we recommend the excellent book of Chillingworth.[59]

Returning to our discussion of limiting behaviour, it is only by scanning through a range of Λ-values, allowing the surface to distort, that we can arrange for the horizontal point of inflexion to appear in typical fashion, as was illustrated in Figure 3.1. This provides the extra variable for our two equations, and the complete phenomenon of the limit point is now structurally stable against any further small perturbation, as was illustrated in Figure 3.3. In similar fashion, to observe a critical equilibrium state where the first three derivatives vanish $(\mathrm{d}V/\mathrm{d}Q = \mathrm{d}^2V/\mathrm{d}Q^2 = \mathrm{d}^3V/\mathrm{d}Q^3 = 0)$, we need a scan through two Λ-parameters for structural stability; this is the case of the cusp catastrophe, underlying the symmetric points of bifurcation discussed earlier.

Structural stability was introduced by the Russians, Andronov and Pontryagin,[60] in the 1930s, but is perhaps better known from the significant developments which took place in the global qualitative theory of dynamical systems in the 1960s, linked with the names of Piexoto, Smale, and Thom, amongst many others.[3-5] In particular, René Thom attaches the utmost importance to the concept, in both mathematical modelling and experimental observation, as we can see from the title of his significant monograph, 'Stabilité structurelle et morphogénèse'.[3] To quote from the translation by D. H. Fowler, Thom states, '... the hypothesis of structural stability of isolated scientific processes is implicit in all scientific observations'.[3]

This emphasis has led to the ideas of *determinacy* and *unfolding*. Let us consider some critical equilibrium state C, any of those experienced by the earlier perfect systems, or possibly something more complex. For the moment, we fix the control parameters of the system, and expand $V(q_i)$ as a Taylor series about C,

$$V = \tfrac{1}{6} V^C_{ijk} q_i q_j q_k + \tfrac{1}{24} V^C_{ijkl} q_i q_j q_k q_l + \text{higher-order terms} \qquad (3.4)$$

Here it is supposed that we have eliminated the passive coordinates, and q_i is thus a set of m incremental active coordinates, measured from C. We see that $V^C_i = 0$ by the first (equilibrium) axiom, and $V^C_{ij} = 0$ because C is an m-fold critical state;[1] some higher-order coefficients may also vanish. The Einstein tensor summation convention is employed, with summations ranging from one to m. The concept of determinacy tells us just which terms are needed for a correct local description of the singularity, and the unfolding rules indicate just how the control parameters are to enter the potential function so as to render the complete phenomenon structurally stable.

Thus the fold catastrophe, or horizontal point of inflexion at C,

$$V = \tfrac{1}{6} V^C_{111} q_1^3 + \text{higher-order terms} \qquad (3.5)$$

requiring just the leading cubic term for determinacy, takes the same local topological form whether or not we remove the higher-order terms by truncation. Moreover, it can be unfolded with the addition of an incremental control parameter λ acting on a linear form of q_1,

$$V = \tfrac{1}{6} V^C_{111} q_1^3 + \lambda V'^C_1 q_1 + \text{higher-order terms} \qquad (3.6)$$

Here the prime denotes partial differentiation with respect to λ. Analysis of this potential function yields the limit point, as we see later.

On the other hand, within a *bifurcational formalism*, the system may exhibit, *a priori*, a control parameter λ acting on the quadratic form. Here we must include a second control parameter ε for structural stability,

$$V = \tfrac{1}{6} V^C_{111} q_1^3 + \tfrac{1}{2} \lambda V'^C_{11} q_1^2 + \varepsilon \dot{V}^C_1 q_1 + \text{higher-order terms} \qquad (3.7)$$

where a dot denotes partial differentiation with respect to ε; this is the asymmetric point of bifurcation of Figure 3.4.

Here we have two quite valid unfoldings of the same singularity, but only the first is described as a *universal unfolding*, since structural stability is achieved with the minimum possible number of control parameters. Nevertheless, the second is also a typical form, and is relevant once we have specified that one of the controls arises in a predetermined way.[6] We note that an excellent and readable account of determinacy and unfolding is given by Poston and Stewart.[5]

To summarize, a singularity will only be observable in the real world if a scan is made through a sufficient number of external control parameters to render the situation non-pathological. The unfolding rules tell us not only how many external parameters are required, but show us exactly how they should enter the potential function. Of equal significance, they inform us that any extra control parameters over and above this particular required number will have no influence on the basic topological form of the singularity. They are thus *essentially ignorable*, and the original problem in $(n + h)$ dimensional space is reduced to a simpler problem in $(m + k)$ dimensional space, where m is the number of *active* coordinates and k is the number of *unfolding* parameters. This reduced space is really an *activity subspace* in which all the essential topology is contained.

Of course, structural stability as we have introduced it here is not a rigorously defined concept, and is open to different interpretations depending on the kind of perturbation envisaged, or even the precise nature of topological change. It might be possible, in the name of rigour, to involve ourselves in lengthy discussions of the appropriate mathematical concepts, with the hope of arriving at precise definitions. However, we have chosen not to do this, first because it would inevitably clash with the non-mathematical nature of the rest of the chapter, and second since, for all the examples considered, no real ambiguity arises here. We note in passing that pure mathematicians are not always in complete accord over the concepts involved, and definitions sometimes vary.

We close this section on structural stability with a practical drawback. It is not always possible to ensure that our mathematical models are structurally stable. For instance, just as we have problems requiring strictly an infinite number of active coordinates (see Section 8.6), so do we have problems needing an infinite number of unfolding parameters. Frequently they are one and the same, as in the notorious buckling of a spherical shell under external pressure, or the cylinder under axial load; here the high degrees of symmetry involved play a very important role. In Chapter 8 we shall discuss such problems in the light of an inherent optimization process; we would refer also to the paper by Poston[61] and the classic text by D'Arcy Thompson[62] for examples of symmetries arising in nature.

We note finally that structurally unstable mathematical models have often been used in the past in buckling analysis, sometimes with a high degree of success. Linear buckling and vibration studies fall into this category, associated as they are with a state of 'neutral' equilibrium, analogous to a ball on a flat table-top. Such formulations do have the virtue of simplicity, but may be unreliable, as we have seen.

3.6 CATASTROPHES OF THOM AND ZEEMAN

Armed with these concepts of structural stability and universal unfolding, Thom was able to classify all the structurally stable forms that can arise in a control space of up to four dimensions. This list of seven elementary catastrophes is presented here in Table 3.1. The proof of the classification theorem is by no means easy, and is not particularly recommended for the non-mathematical reader, who may be best advised simply to accept the result. A complete proof can be found in Zeeman's book,[4] and a readable outline in the book by Poston and Stewart.[5]

Table 3.1 The list of seven elementary catastrophes. This shows the seven forms of instability that can arise in a conservative system with no more than four control parameters. The name of the catastrophe is followed by the simplest potential polynomial that can generate it. The q_i represent the internal state variables (generalized coordinates) and the λ^j the external control parameters. The right-hand column shows some features of a finer bifurcational classification. Notice that all potential functions are linear in the control parameters λ^j

$m = 1$	Fold	$q_1^3 + \lambda^1 q_1$	{ Limit point { Asymmetric
	Cusp	$q_1^4 + \lambda^2 q_1^2 + \lambda^1 q_1$	{ Stable-symmetric { Unstable-symmetric { Cut-off point
	Swallowtail	$q_1^5 + \lambda^3 q_1^3 + \lambda^2 q_1^2 + \lambda^1 q_1$	
	Butterfly	$q_1^6 + \lambda^4 q_1^4 + \lambda^3 q_1^3 + \lambda^2 q_1^2 + \lambda^1 q_1$	
$m = 2$	Hyperbolic umbilic	$q_2^3 + q_1^3 + \lambda^1 q_2 q_1 - \lambda^2 q_2 - \lambda^3 q_1$	{ Monoclinic { Homeoclinic { Hill-top branch
	Elliptic umbilic	$q_2^3 - 3q_2 q_1^2 + \lambda^1 (q_2^2 + q_1^2) - \lambda^2 q_2 - \lambda^3 q_1$	Anticlinal
	Parabolic umbilic	$q_2^2 q_1 + q_1^4 + \lambda^1 q_2^2 + \lambda^2 q_1^2 - \lambda^3 q_2 - \lambda^4 q_1$	Paraclinal

We note that the word 'elementary' here refers to the fact that the system is governed by a potential function, not to the limitations of four dimensions of control, as is sometimes supposed. Thus it is possible to find elementary catastrophes lying outside this list; we shall consider later one example of some significance, the double-cusp catastrophe, which requires at least seven parameters for a universal unfolding. The analytical complexities of such a problem are considerable, far more difficult than anything to be attempted here. We shall consider a full unfolding only for the phenomena on Thom's list.

Looking at this list in a little more detail, we see that the first four have a single active coordinate; these are sometimes referred to collectively as the cuspoid catastrophes. The remaining three, the umbilics, have two active coordinates, so the simplest three-fold ($m = 3$) singularity is not included. This may come as something of a surprise to some buckling analysts, since it might be supposed that the four controls of load plus three imperfections, one in each buckling mode, would be fully sufficient to understand an $m = 3$ singularity.

The table also includes the simplest possible (determinate) potential function for each singularity, with appropriate control parameters for a universal unfolding included. On the right-hand side we outline some of the more detailed classification obtained under a bifurcational formalism. This has a somewhat finer mesh than Thom's, and we note that it corresponds more closely to the refined viewpoints of Wassermann,[63,64] and Golubitsky and Schaeffer,[6,65,66] who also consider one control parameter as being, in some way, fundamentally different from the rest. In the following chapters we shall be looking in detail at most of the phenomena listed here.

Now, we have restricted attention to $k \leq 4$, where k is the dimension of the control space, but Thom's theorem is often stated in five dimensions, and here another four catastrophes emerge;[4] for $k = 6$ an infinite number of forms are possible, and there is thus no complete classification. Thom's original restriction to four dimensions arose because of his interest in biology, where a control space can be sometimes related to space-time. With imperfections as controls we may at times need to go beyond $k = 5$, and it is interesting here to enquire just what takes us so far from typicality. Frequently this is an inherent optimization process, causing a high-order singularity in the response of the perfect system, which, of course, must be included in the mathematical model although it can never be experienced by a real physical system.

It is the classification theorem which lies at the heart of the contribution now known as *elementary catastrophe theory*. The theorem can be put to a surprising variety of uses. First, we have the contribution of Thom himself, with the emphasis mainly on morphogenesis in biology.[3] Second, we must mention the large number of applications is different disciplines due to Zeeman[4] and also Poston and Stewart,[5] who concentrate largely on the physical sciences. An ingenious *predictive* use of the theorem in light caustics by Berry[25] is fully confirmed by experiment.

3.7 BIFURCATIONS OF GOLUBITSKY AND SCHAEFFER

The unfolded singularities of Thom and Zeeman can be seen as standing with an arbitrary orientation above a virgin control space, as shown by the un-parametrized equilibrium surface of Figure 3.10(a). This is the *catastrophe theory* view, illustrated here by the dual cusp. Notice that the stability boundary projects into a two-thirds power-law cusp in the two-dimensional control space of the base plane.

By contrast, in a *bifurcational* view we suppose that a primary control parameter Λ is specified *a priori* by the problem in hand. This means, effectively, that the control space is parametrized *a priori*, so that the variation of Λ with $\varepsilon = 0$ specifies an evolutionary *route*[13] through the control space. An equivalent statement is that $\lambda = \Lambda - \Lambda^C$ and ε might already act on particular powers of q in a local Taylor expansion.

A particular parametrization of the dual cusp is shown in Figure 3.10(b). Here we have drawn the Λ axis down the axis of the cusp, and constant-ε slices are drawn on the surface to reveal the equilibrium paths of the unstable-symmetric point of bifurcation. The same parametrization is also shown in Figure 4.9.

Now the situation we have drawn here is clearly *not typical* because of the precise coincidence of the Λ axis with the cusp axis. So an *a priori* parametrization clearly demands a new look at the concept of structural stability.

The need for a finer *bifurcational* classification of instability phenomena was first noted by Thompson and Hunt,[45] and the structurally stable bifurcations have now been analysed in a powerful paper by Golubitsky and Schaeffer.[6] These workers have dropped the need for a system to be conservative, and operate just on the equilibrium equations, corresponding to $V_i(Q_j, \Lambda^k) = 0$, which in their work need not have the property that $V_{ij} = V_{ji}$. This being so, they only look at

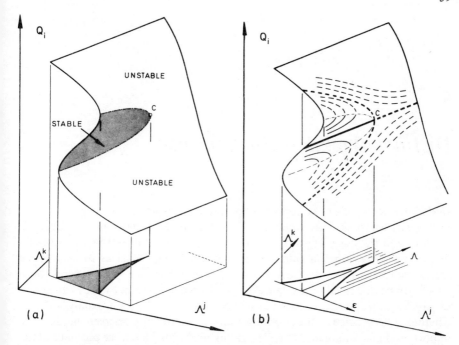

Figure 3.10 The catastrophe theory view (a) and the bifurcational view (b) of a cusp corresponding to an unstable-symmetric point of bifurcation

the form of the equilibrium paths, and make no statements about *stability*: such statements follow in a natural way from the equilibrium forms of a conservative system, but not necessarily from the equilibrium forms of a non-conservative system.

Among other more unusual structurally stable bifurcations they identify our asymmetric point of bifurcation and our cut-off point (which they call a non-degenerate hysteresis point). They also show that for typicality the symmetric bifurcation of Figure 3.10(b) needs in its V function not only the normal imperfection term $\varepsilon_1 q$ but also a tilting term $\varepsilon_2 q^3$. This latter, on its own, tilts the underlying cusp and converts the symmetric point of bifurcation locally into an asymmetric bifurcation.[67]

We shall be looking more closely at the structurally stable catastrophes and bifurcations in the following chapters.

An excellent modern topological review of the Thom–Zeeman and Golubitsky–Schaeffer viewpoints is given by Chillingworth.[68] Qualitative methods in modern bifurcation theory are outlined by Marsden[69] and Holmes and Marsden.[70] The latter paper covers the dynamic bifurcations of non-conservative systems, and the chaotic motions of strange attractors: these attractors, with their cascades of period-doubling bifurcations, must now be expected in the resonance of conservative buckling systems.[71,72] The analysis of structures within a catastrophe theory framework is presented recently by Niwa *et al.*,[73] who employ discretization and modal transforms.

4

Distinct Buckling Phenomena

In this chapter we are to be concerned purely with *distinct* buckling phenomena, or those with a single active generalized coordinate Q_1. The two most important, the fold and the cusp, are dealt with in some depth, and we shall also briefly discuss two less common singularities. By restricting ourselves to $m = 1$ (one active coordinate) we can treat all one-degree-of-freedom systems, and those which can be described by a single effective degree of freedom by the elimination of passive coordinates, as outlined in Chapter 3 and described analytically in Chapter 7. We cannot discuss, for instance, interaction between modes, the important ingredient of many shell buckling problems; this is the province of the umbilics and higher-order singularities, and is treated in Chapter 6.

The phenomena are fully illustrated with simple mechanical spring models wherever possible, and a complete post-buckling study of the Euler strut is also included. We explore in particular the significance of one of the controls, in buckling studies the loading parameter Λ, being in some way different from the rest, the imperfections, denying the freedom to mix the two types in transformations; this, as we have seen in Section 3.7, is the essence of the so-called *bifurcational formalism*.[6]

4.1 THE FOLD SINGULARITY

The fold embraces two of the distinct critical points introduced in the previous chapter, the limit point and the asymmetric point of bifurcation. Of these, the limit point is the most general, and is the only structurally stable phenomenon that can arise in a one-dimensional control space. It is best illustrated by a simple example, so we start with a single-degree-of-freedom arch analysis. Unlike the arch model of our earlier book,[1] the non-linear response is traced in its entirety without resort to assumptions of moderate deflection. It exhibits two limit points which arise in a common form, complying with an underlying symmetry.

Tied arch model

Consider the shallow arch of Figure 4.1, comprising two rigid links of length L, pinned at an apex and tied by an extensional spring of stiffness k as shown. The

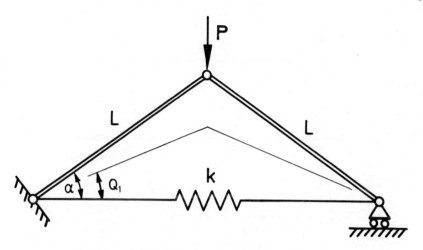

Figure 4.1 A single-degree-of-freedom tied arch

spring is capable of resisting both tension and compression. The initial angle of the links to the horizontal is α, and since only symmetric deformations are possible the system has one degree of freedom which we take as the variable angle Q_1. The central pin is loaded by a dead vertical force of magnitude P.

The original length of the spring is $2L \cos \alpha$ and the subsequent length is $2L \cos Q_1$. Thus the strain energy becomes

$$U(Q_1) = \tfrac{1}{2}k(2L\cos Q_1 - 2L\cos\alpha)^2 = 2kL^2(\cos Q_1 - \cos\alpha)^2 \qquad (4.1)$$

The corresponding deflection of the dead load P is simply

$$\mathscr{E}(Q_1) = L(\sin\alpha - \sin Q_1) \qquad (4.2)$$

The potential energy of the load is $-P\mathscr{E}(Q_1)$, so the total potential energy of the system (structure plus load) becomes

$$\begin{aligned} V(Q_1, P) &= U(Q_1) - P\mathscr{E}(Q_1) \\ &= 2kL^2(\cos Q_1 - \cos\alpha)^2 - PL(\sin\alpha - \sin Q_1) \end{aligned} \qquad (4.3)$$

The condition for equilibrium is simply

$$\frac{\partial V}{\partial Q_1} \equiv V_1 = 4kL^2(\cos Q_1 - \cos\alpha)(-\sin Q_1) + PL\cos Q_1 = 0 \qquad (4.4)$$

and we have the equilibrium solution

$$P(Q_1) = 4kL(\cos Q_1 - \cos\alpha)\tan Q_1 \qquad (4.5)$$

giving the single equilibrium path shown in Figure 4.2. We find that P is stationary with respect to Q_1 when $\cos^3 Q_1 = \cos\alpha$, and zero when $Q_1 = 0$ or $\pm\alpha$.

To study the stability of the equilibrium states we first form the second derivative,

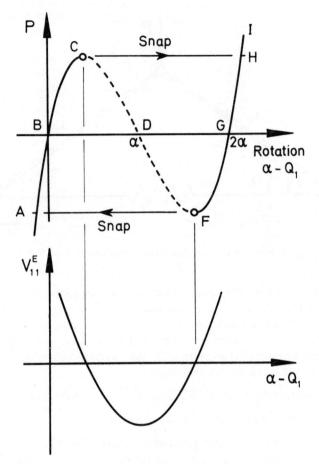

Figure 4.2 Equilibrium path of the tied arch, and variation of V^E_{11} indicating regions of stability and instability

$$\frac{\partial^2 V}{\partial Q_1^2} \equiv V_{11} = 4kL^2(\sin^2 Q_1 - \cos^2 Q_1 + \cos Q_1 \cos \alpha) - PL\sin Q_1 \quad (4.6)$$

and evaluating at each point E on the equilibrium path by substituting for P from equation (4.5) we obtain

$$V^E_{11} = 4kL^2 \frac{\cos \alpha - \cos^3 Q_1}{\cos Q_1} \quad (4.7)$$

which varies with Q_1 as shown in Figure 4.2. Considering a Taylor expansion about E with P constant, $V^E_1 = 0$, so if V^E_{11} is positive we have a local minimum of V and the equilibrium state is stable, while if V^E_{11} is negative we have a local maximum of V and the equilibrium state is unstable. The stability of the path thus changes at the critical states C and F, as shown in Figure 4.2, where, as before, stable equilibrium states are represented by a solid line and unstable equilibrium states are represented by a broken line. We see that the two unloaded equilibrium states represented by $Q_1 = \pm \alpha$ are stable while the unloaded equilibrium state represented by $Q_1 = 0$ is unstable, as we might expect.

It remains to enquire whether the critical equilibria, C and F of Figure 4.2, are themselves stable or unstable. In these states we have $V_1^E = V_{11}^E = 0$, and to examine the local form of V we must clearly examine the third derivative. This is readily obtained as

$$V_{111}^E = 6kL^2 \sin 2Q_1 \qquad (4.8)$$

and we see that it is non-zero at the critical states C and F. Thus considering a Taylor expansion of V about these states we see that V is locally cubic, giving a horizontal point of inflexion; so V is not a minimum, and the two critical equilibrium states are themselves unstable.

These critical states C and F, where a single smooth equilibrium path reaches a maximum (or minimum) in load-coordinate space, are known as *limit points*. For a single-degree-of-freedom system such a point must clearly herald a change from stability to instability on the path, but for $n \neq 1$ this is not necessarily the case, since instabilities can arise with respect to different critical coordinates in turn; then all we can say is that the *degree of instability* changes by one at a distinct limit point.[18]

A loading sequence can now be traced on Figure 4.2. Starting from the unloaded state B the arch will follow the stable rising path BC as the load is increased. On reaching the unstable limit point C the arch will snap dynamically under the prescribed value of the dead load P and will exhibit large-amplitude vibrations about the stable equilibrium state H. In the presence of a little positive–definite viscous damping the energy released in the displacement from C to H (the equivalent of the area $CDFGH$ on a plot of load against its corresponding displacement) will eventually be dissipated and the system will come to rest in state H. On further loading the arch, now inverted, will follow the stable rising path HI, as indicated.

On unloading a similar sequence is followed. Thus as P is decreased the arch will remain inverted along the stable path $IHGF$ and will then snap dynamically from F to A, an increase in P then being necessary to bring the arch back to the unloaded state B.

We see that the arch has followed a hysteresis loop $CHFA$, the corresponding energy having been dissipated in the two dynamic snaps. The region CF of the equilibrium path corresponding to unstable states of equilibrium is never encountered in a normal loading sequence, but the equilibrium states do, of course, exist, and can be encountered experimentally if a rigid loading device is used.[18]

4.2 SNAP-BUCKLING AT A LIMIT POINT

This simple model, and the general discussion of the previous chapter, paint a fairly comprehensive picture of limiting behaviour, and it is not necessary to say very much more about this relatively simple phenomenon. Analysis in a general theory context, employing a set of fixed coordinates at the critical point, can be found in our earlier book,[1] and is not worth repeating in isolation here. We merely note that, after the elimination of passive coordinates, the limit point is described by the potential function

$$V = \tfrac{1}{6}V^C_{111}q^3_1 + V'^C_1\lambda q_1 + \text{higher-order terms} \tag{4.9}$$

and stress in particular that the incremental control parameter $\lambda = \Lambda - \Lambda^C$, where C denotes evaluation at the critical point itself, appears in a term that is linear in $q_1 = Q_1 - Q^C_1$, the incremental active coordinate. Notice that q_1 is here measured always from the fixed critical point itself, and is not a sliding coordinate as in our later discussions of branching points.

We also note once again that this most common form of the fold catastrophe is the only structurally stable phenomenon that can arise in a control space of just one dimension, so any (real and therefore imperfect) structure under the action of a single load must fail at a limit point. But the behaviour of some underlying perfect system, complete with a higher-order singularity, may nevertheless be crucial; we return to this in the discussion of symmetry and optimization in Chapter 8.

Finally, for a number of theorems concerning the general limit point (seen as one of a succession of folds), together with a full discussion of the difference between dead and rigid loading of structures, we refer to a recent Royal Society publication.[18]

4.3 ASYMMETRIC POINT OF BIFURCATION

The fold catastrophe takes on a new significance once we allow a second dimension of control, which in the physical sciences is often provided by an imperfection parameter. This extends the fold into a third dimension, and we arrive at a folded surface of equilibrium states as shown in the top-left diagram of Figure 4.3; here we have plotted the single active coordinate vertically, the base plane becoming the control space, following the convention of general topological studies. We suppose that the lower surface represents stable equilibria, shown shaded in the top-left diagram, and the upper surface represents unstable equilibria. The regions of stability and instability are separated by a line of critical fold points, giving, by projection, a smooth failure locus in control space as shown.

It is interesting to observe that, for states of non-equilibrium lying off the surface, the three-dimensional space is separated by the surface into two distinct regions of opposite vertical flow, indicating the fast dynamic response of the system. This is shown on the top-left diagram by the broad arrowheads.

A bifurcational formalism, as discussed in Section 3.7, takes parallel vertical slices through such an equilibrium surface, the intersections of these planes with the surface giving the equilibrium paths at different but constant values of one of the controls; thus equilibrium paths are no more than contours of a general equilibrium surface. Normally we would expect that a single slice through the

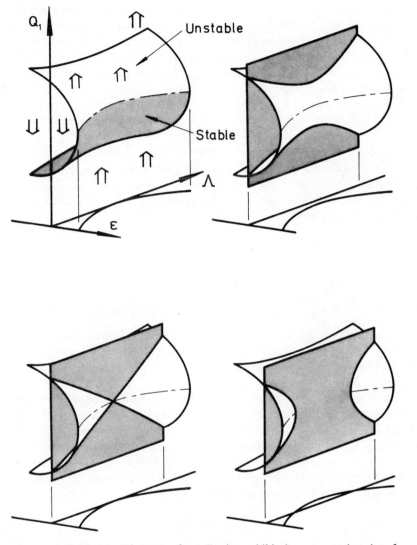

Figure 4.3 Folded equilibrium surface, sliced to exhibit the asymmetric point of bifurcation

surface of Figure 4.3 would give the typical form of the limit point, and that the family of parallel cuts would generate further limit points because of its structural stability; thus we would have a locally linear imperfection-sensitivity in control space, as discussed in the previous chapter and shown in Figure 3.3.

But this is not the only way in which the surface of Figure 4.3 can be sliced, as we show in the remaining three diagrams. Here, because of the special orientation of the Λ-direction, the intersections of the vertical planes and the surface yield the equilibrium paths of the asymmetric point of bifurcation, introduced in the previous chapter and shown in Figure 3.4. One of the slices, shown at the bottom left, gives the two intersecting equilibrium paths which exhibit an *exchange of stability*, as first discussed by Poincaré;[39] we are free to identify this special case with the response of a *perfect system*, giving a Λ-axis which just touches the failure locus in control space and a locally parabolic form of imperfection-sensitivity. The two right-hand diagrams show the typical responses of imperfect systems.

This pattern of equilibrium paths, although less likely to arise in general than limiting behaviour, is nevertheless a structurally stable phenomenon once we have introduced the second element of control; for a very clear discussion of this in a mathematical context we recommend the comprehensive article by Golubitsky and Schaeffer.[6] It is clear from Figure 4.3 that an asymmetric point of bifurcation can only arise where the equilibrium surface has a negative Gaussian curvature (saddle-like appearance). It turns up in a number of different structural problems, including frames, plates, and shells,[1] and is of some significance in fluid mechanics applications.[32] It is also an important ingredient of the umbilic catastrophes, as we shall see in Chapter 6.

After the elimination of passive coordinates the asymmetric point of bifurcation is generated by the Taylor expansion of the potential function

$$V = \tfrac{1}{6}V^C_{111}q_1^3 + \tfrac{1}{2}V'^C_{11}\lambda q_1^2 + \dot{V}^C_1 \varepsilon q_1 + \text{higher-order terms} \qquad (4.10)$$

and we note now the appearance of the distinctive *bifurcational parameter* λ (normally a load) acting on the quadratic term q_1^2; this is usually given *a priori* by the problem in hand, and structural stability now calls[6] for the extra control ε (an imperfection) acting on the linear form q_1.

In many applications of the physical sciences one of the equilibrium paths of the perfect system (the primary or fundamental path) involves a simple, perhaps trivial, solution to the governing equations, and the problem is essentially one of seeking information concerning the secondary or post-buckling path; this is the context of the general bifurcation analysis of Chapter 7. Thus, or by similar earlier techniques,[1] we can write down an exact expression for the (non-zero) initial slope of the secondary equilibrium path

$$\frac{d\Lambda}{dq_1}\bigg|^C = -\frac{V_{111}}{2V'_{11}}\bigg|^C \qquad (4.11)$$

We see that this same relation can be obtained by truncating the potential function equation (4.10) after the leading two terms and setting $V_1 = 0$ for equilibrium, to give the local form of the perfect system response.

Similarly, the leading half power law term of the asymptotic equation of imperfection-sensitivity is given by

$$\Lambda = \Lambda^C \pm (V^C_{111})^{1/2}\frac{(2\dot{V}^C_1 \varepsilon)^{1/2}}{V'^C_{11}} \qquad (4.12)$$

a result that again can be obtained by truncating equation (4.10), this time after the first three terms, and setting $V_1 = 0$ for equilibrium and $V_{11} = 0$ for criticality.

We note that these expressions are only valid when V is written in terms of a 'sliding' incremental active coordinate q_1, measured always from the fundamental path itself, not just from the critical point. Higher-order terms can be obtained by extending the perturbation schemes, and are given explicitly in our book.[1]

A simple asymmetric model

We illustrate the asymmetric point of bifurcation again via a simple model. Consider the structural system shown in Figure 4.4 comprising a rigid link of length L, supported by a linear spring of stiffness k in both tension and compression and inclined initially at 45 degrees. The structure is loaded by a vertical dead load of magnitude P. To generate a family of imperfect systems we suppose that the load is offset by a distance εL, and to show the effect of a second imperfection we shall assume additionally that the spring is initially too long by $\gamma \varepsilon L \sqrt{2}$, where γ is an arbitrary constant. The side-sway of the structure is denoted by $Q_1 L$, the single degree of freedom being the non-dimensional generalized coordinate Q_1.

The strain energy of the system is given by

$$U = kL^2[(1 + Q_1)^{1/2} - 1 - \gamma\varepsilon]^2 \tag{4.13}$$

and the corresponding deflection of the load P becomes

$$\mathscr{E} = L[1 + Q_1\varepsilon - (1 - Q_1^2)^{1/2}] \tag{4.14}$$

so the total potential energy of the system can be written as

$$V(Q_1, P, \varepsilon) = kL^2[(1 + Q_1)^{1/2} - 1 - \gamma\varepsilon]^2 - PL[1 + Q_1\varepsilon - (1 - Q_1^2)^{1/2}] \tag{4.15}$$

Let us consider first the behaviour of the perfect system by setting $\varepsilon = 0$ in the potential energy expression, thereby ensuring that both the imperfections are

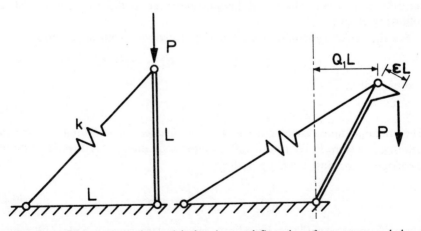

Figure 4.4 The asymmetric model showing undeflected perfect system and deflected imperfect system

absent. The equilibrium condition now gives

$$V_1(Q_1, P) = kL^2[1 - (1 + Q_1)^{-1/2}] - PLQ_1(1 - Q_1^2)^{-1/2} = 0 \qquad (4.16)$$

and we see that we have a trivial fundamental solution F given by $Q_1 = 0$. We investigate the stability of this by the second differentiation

$$V_{11}(Q_1, P) = \frac{kL^2}{2}(1 + Q_1)^{-3/2} - PL(1 - Q_1^2)^{-1/2} - PLQ_1^2(1 - Q_1^2)^{-3/2} \qquad (4.17)$$

giving, by evaluation on F, the stability coefficient

$$V_{11}^F = \frac{kL^2}{2} - PL \qquad (4.18)$$

Clearly, we have a critical point at $Q_1 = 0$, $P^C = kL/2$, the fundamental path being stable below and unstable above this point (with P plotted vertically). The critical equilibrium state itself can be seen to be unstable by a further differentiation

$$V_{111} = -\tfrac{3}{4}kL^2(1 + Q_1)^{-5/2} - 3PQ_1(1 - Q_1^2)^{-3/2} - 3PLQ_1^3(1 - Q_1^2)^{-5/2} \qquad (4.19)$$

giving

$$V_{111}^C = -\tfrac{3}{4}kL^2 \qquad (4.20)$$

It now remains to consider the post-buckling behaviour of this perfect system.

In the original study,[1] closed-form solutions were not sought, and the model was used to introduce perturbation methods of analysis; appropriate first-order equations were derived, both implicitly and explicitly, and solved for the slope of the post-buckling path and the locally dominant term of the imperfection-sensitivity equation. Here we shall merely use the general theory forms, given in the previous section, to obtain the same results directly. For a specific introduction to perturbation techniques we refer to the arch analysis of the following chapter.

For the perfect system, we must first differentiate V_{11} with respect to P to give

$$V_{11}' = -L(1 - Q_1^2)^{-1/2} - LQ_1^2(1 - Q_1^2)^{-3/2} \qquad (4.21)$$

and evaluating at C,

$$V_{11}'^C = -L \qquad (4.22)$$

The fundamental path being trivial, Q_1 is an *incremental* coordinate measuring deformation from F. We can therefore employ equation (4.11) directly and write the slope of the post-buckling path as

$$\left.\frac{\mathrm{d}P}{\mathrm{d}Q_1}\right|^C = -\left.\frac{V_{111}}{2V_{11}'}\right|^C = -\tfrac{3}{8}kL \qquad (4.23)$$

The post-buckling path thus has a finite, but non-zero, slope on a plot of P against Q_1.

But if we plot the load against its corresponding deflection this is no longer the case, and the two paths, fundamental and post-buckling, just touch at C. We can show this to be true by inverting $\mathscr{E}(Q_1)$ to give

$$Q_1 = \pm \left[\frac{\mathscr{E}}{L} \left(2 - \frac{\mathscr{E}}{L} \right) \right]^{1/2} \tag{4.24}$$

and substituting this into the truncated Taylor expansion $P(Q_1)$ to give the first-order result

$$P = P^C \pm \frac{3\sqrt{(2)}}{8} kL \left(\frac{\mathscr{E}}{L} \right)^{1/2} \tag{4.25}$$

implying a locally parabolic variation at C. The distinction between the forms of load to generalized coordinate and load to corresponding deflection plots for the perfect system is common to all points of bifurcation, and is associated with the fact that, unlike in the limit point, the load does no first-order work as the system moves through its buckling displacement.[1]

Moving on now to the imperfection-sensitivity, we reintroduce the imperfections into the potential function and perform a similar series of differentiations. The derivatives already evaluated at the critical point of the perfect system remain unchanged, and differentiating the general form for V_1 with respect to ε we are left with

$$\dot{V}_1 = -kL^2\gamma(1 + Q_1)^{-1/2} - PL \tag{4.26}$$

where a dot denotes partial differentiation with respect to ε, giving

$$\dot{V}_1^c = -\frac{kL^2}{2}(1 + 2\gamma) \tag{4.27}$$

Substituting the known derivatives into equation (4.12), we obtain the initial half power law term of the full asymptotic equation of imperfection-sensitivity as

$$P = \frac{kL}{2}\{1 \pm [3(1 + 2\gamma)\varepsilon]^{1/2}\} \tag{4.28}$$

The family of equilibrium paths and the imperfection-sensitivity are shown schematically in Figure 4.5. The trivial fundamental path of the perfect system coincides with the P-axis in the left-hand diagram, and intersects the post-buckling path at the critical point C. We note that it is a general result, common to all asymmetric bifurcations, that the initial slope of the locus of critical limit points for imperfect systems, SS, is numerically equal to twice the initial slope of the post-buckling path of the perfect system. The right-hand diagram shows the initially parabolic imperfection-sensitivity. The locally valid results obtained above are shown contaminated by higher-order effects as we emerge from C, curving the post-buckling path and distorting the parabolic form of the imperfection-sensitivity.

The appearance of γ in the imperfection-sensitivity result indicates clearly the effect of adding a second imperfection. In general, it can lead to no new

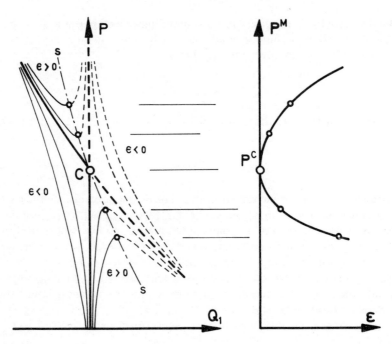

Figure 4.5 Schematic representation of the response of perfect and imperfect systems (for $1 + 2\gamma > 0$)

topological features, but if $1 + 2\gamma = 0$, the two imperfections just cancel one another, and a bifurcation of equilibrium path arises in the response of all systems; this would not constitute a structurally stable unfolding of the singularity.

An asymmetric frame

As a final illustration of the asymmetric point of bifurcation in a structural problem we shall briefly discuss the asymmetric frame shown in Figure 4.6. Here we have two identical members of length l, pinned to rigid supports and rigidly joined together at right-angles, as shown. The frame is loaded by the dead vertical load P which retains its magnitude and direction as the system deflects. The members of the frame are assumed to be axially inextensional and the bending stiffness of each bar is the same.

This system was first studied experimentally by Roorda,[53] and subsequently modelled as a continuum by Koiter,[74] who arrived at expressions for critical load, slope of the post-buckling path, and leading term of the imperfection-sensitivity relationship. In our book[1] we present a discrete coordinate, finite-element study of the same frame, to the same order of complexity. It is found to require very little extra effort over and above that demanded by a linear eigenvalue analysis; only quadratic terms of strain energy and end-shortening functions of individual

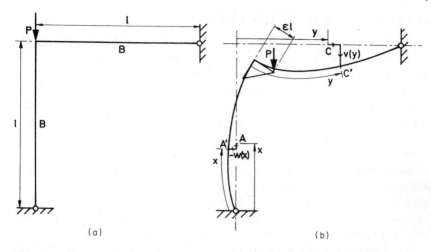

Figure 4.6 The two-bar frame showing (a) undeflected perfect system and (b) deflected imperfect system

elements are needed, the cubic term of potential energy arising from a combination of these via a non-linear compatibility relationship at the knee of the frame. Convergence to Koiter's exact values with increasing number of degrees of freedom is very rapid, the results all lying within 3% for as few as two elements per member. The comparison between theory and experiment is found to be most satisfactory, shown here in Figure 4.7.

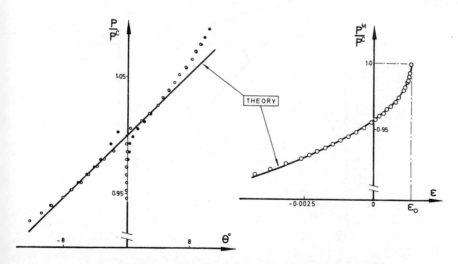

Figure 4.7 Comparison between finite element results and the experiments of Roorda[53]

72

4.4 ROUTES THROUGH THE FOLD

We finally introduce a concept which is, to some extent, self-evident for the fold, but is of considerable help later in examining and classifying higher-order phenomena. Let us suppose that we have three dimensions of control, but the system can experience only the fold catastrophe and nothing of higher order. The failure locus is then, in general, a smooth curved sheet in control space, as shown in Figure 4.8. The sheet divides the control space into a region having two equilibrium solutions, one stable, denoted by a solid circle, and one unstable, denoted by an open circle, and a region having no equilibrium solutions.

Considering a general curvilinear coordinate system, the variation of a single control such as a structural load generates a *route* through the space; in practice this route must start from a point associated with a stable state of equilibrium. Two possibilities are shown. Route *A* passes directly through the critical boundary and gives the limit point of the upper-left diagram. In contrast, route *B* touches the critical boundary, and generates the asymmetric point of bifurcation of the lower-left diagram.

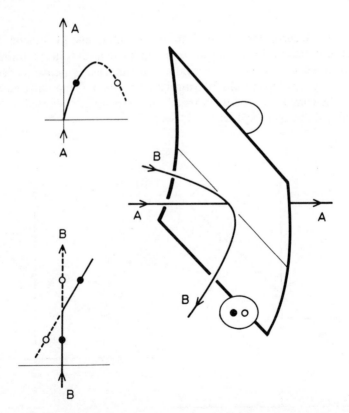

Figure 4.8 Routes in control space through the fold, giving the limit point and the asymmetric point of bifurcation

4.5 THE CUSP SINGULARITY

Of all the singularities on Thom's list, the cusp has received the most attention in the literature; we note in particular that Zeeman has applied it to a broad spectrum of disciplines.[4] Ignoring for the moment our bifurcational formalism, it is clear why this should be the case. Apart from the fold, which, taking its general form of the limit point, can be regarded as trivial, the cusp is not only the simplest possible singularity but also the most common, arising as it does in the sub-strata of all higher-order singularities. Moreover, it exhibits certain features—hysteresis, divergence, and history-dependence, for example—that often seem to occur together in nature but are missing from limiting behaviour. But, perhaps most significantly, it is relatively easy to visualize, requiring, before the introduction of the bifurcational formalism, two dimensions of control for structural stability.

The geometry of the cusp

To interpret the geometry of the cusp catastrophe we start with the top-left diagram of Figure 4.9. Here, like the folded surface of Figure 4.3, we have the active coordinate Q_1 plotted vertically, the base plane being the two-dimensional control space. We again have a smooth surface in this three-dimensional space, with a single line of critical (generally fold) points lying on it, shown as the broken line. For non-equilibria the surface separates the full three-dimensional space into two regions of opposite fast dynamic response (the broad arrowheads), while the line of critical states separates the surface into two regions of stable and unstable equilibrium, as before.

But otherwise the geometry of the surface is very different from the earlier folded surface. It has a gradually changing section, from the S-shape at the front to the monotonic form at the back, and thus exhibits a single, higher-order critical state marking the exact transition. This is the cusp point itself, and here not only does the surface have a vertical tangent but so does the line of critical states. Projecting the latter into control space we have the shape of a two-thirds power law cusp. We see that this locus separates the control space into two regions, points lying outside the cusp exhibiting just one possible equilibrium state, while those inside have three.

We note that this geometry can take two entirely opposite forms. The situation shown at the top-left of Figure 4.9 is usually referred to as the *dual-cusp* catastrophe. For the cusp catastrophe itself we must interchange the stable and unstable regions, and reverse the direction of the fast dynamic response shown by the broad arrowheads.

This surface, and, of course, the fold, are the only smooth structurally stable phenomena that can arise with just two dimensions of control; we note that this holds in general singularity theory, without the restriction of a potential function—a strong result due to Witney which has no immediate parallel in higher dimensions. Its structural stability can be neatly illustrated by a small random tilt of the surface with respect to our Euclidean reference frame. This will

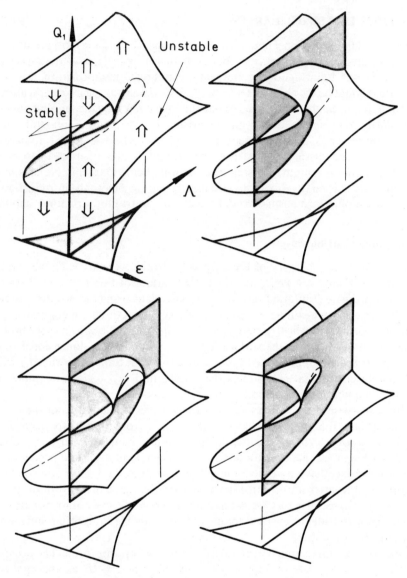

Figure 4.9 Equilibrium surface of a dual cusp catastrophe, sliced to exhibit
the unstable-symmetric point of bifurcation

shift the cusp point, and the line of folds, slightly on the surface, but otherwise
everything will remain qualitatively the same; we might say that the cusp point
itself only takes on a special significance once we define the direction of Q_1.
Similarly, by viewing the surface from an oblique angle as we do in Figure 4.9, we
must see a cusp, although some of it is obscured by the upper part of the surface;

the position of this visible cusp is, of course, well displaced from the original, and would only coincide with it if we viewed the surface from on top.

Returning to the bifurcational formalism, we again slice up the control space, so that the intersections of a set of parallel vertical planes with the surface form the equilibrium paths; we have seen this earlier in Figure 3.10. With Λ orientated along the axis of symmetry of the cusp, as shown in the top-left of Figure 4.9, this gives the typical slices of the other three diagrams. We can recognize the familiar 'pitchfork' pattern of equilibria of a symmetric point of bifurcation in the bottom-left diagram. Associating this with the response of some perfect system as before, typical forms for imperfect systems are as shown in the right-hand diagrams. The stability of the paths is not indicated here, but from the top-left it is clear that we have an unstable-symmetric point of bifurcation.

We see that the symmetric point of bifurcation comes from a very special slice of the surface, with Λ directed along the cusp axis. More typically, we would expect the cusp to be tilted with respect to the Λ-direction as mentioned in Section 3.7, and the cusp point itself then arises not as a bifurcation of equilibrium paths but as a *cut-off point*. We shall return later to more detailed discussions of routes through the cusp, cut-off points, and the tilted cusp.

But the lack of typicality of the slice does imply that, for structural stability, a symmetric point of bifurcation needs a third control parameter, to tilt the cusp; this has been proved in a mathematical context by Golubitsky and Schaeffer,[6] and is briefly discussed in Section 3.7. The third parameter could be seen as acting on a cubic form of potential energy, and in structural mechanics applications, for example, might appear as an extra imperfection; with just a casual examination of the orders involved we can, however, suggest that its physical significance would be minimal when compared with the imperfection acting on a linear form, a conclusion that is reinforced by the experimental evidence.[54] So here, we have a situation where structural instability seems to carry few dramatic consequences. However, this would not be the case if we found a primary control, such as a load, acting on the cubic form.

4.6 STABLE-SYMMETRIC POINT OF BIFURCATION

After the elimination of passive coordinates both symmetric points of bifurcation are generated by the Taylor expansion of the potential function

$$V = \tfrac{1}{24}V^C_{1111}q_1^4 + \tfrac{1}{2}V'^C_{11}\lambda q_1^2 + \dot{V}^C_1 \varepsilon q_1 + \text{higher-order terms} \qquad (4.29)$$

where the incremental bifurcational parameter λ and the imperfection ε can be seen to act on quadratic and linear terms in q_1, respectively, as before: we have omitted the third control parameter of Golubitsky and Schaeffer,[6] which would otherwise be found acting on q_1^3. For the stable-symmetric point of bifurcation, shown earlier in Figures 3.5 and 3.7, V^C_{1111} is positive, and we are slicing up the cusp catastrophe, the inverse of Figure 4.9, which shows the dual cusp.

In physical science applications we usually have a single-valued fundamental equilibrium path which loses its stability at the critical point, and we can identify this as the central path of the pitchfork in the bottom-left diagram of Figure 4.9, with the Λ-direction reversed. Using the bifurcation analysis of Chapter 7, or earlier techniques,[1] we can write down the initial curvature of the secondary equilibrium path—the two outside prongs of the pitchfork—as

$$\frac{d^2 \Lambda}{dq_1^2}\bigg|^C = -\frac{V_{1111}}{3V'_{11}}\bigg|^C \tag{4.30}$$

We note again that this expression is only valid with V written in terms of a sliding incremental generalized coordinate q_1. Also, the same relation can alternatively be obtained by truncating V directly, as before.

The imperfection-sensitivity expression is not given explicitly here, since it has little physical significance for systems exhibiting a stable-symmetric point of bifurcation. Higher-order terms in a Taylor expansion of the secondary path can be obtained by extending the perturbation analysis, and are given explicitly in our book.[1]

Post-buckling of the Euler strut

Chapter 2 includes a large-deflection inextensional beam and strut formulation, giving expanded forms for strain energy U and end-shortening $\mathscr{E}$ in terms of derivatives of the deflection function $w(x)$. The leading terms are then used in small-amplitude vibration and strut-buckling eigenvalue analyses. We here use the second terms of the series, to illustrate the stable-symmetric bifurcation of the pin-ended strut, and determine the curvature of the post-buckling path.

As before, the deflection function $w(x)$ of Figure 2.1 is written as a modal expansion,

$$w(x) = \sum_{i=1}^{\infty} q_i \sin \frac{i\pi x}{L} \tag{4.31}$$

the generalized coordinates q_i now no longer being functions of time in this static post-buckling analysis. Substituting into the strain energy U of equation (2.5) and end-shortening $\mathscr{E}$ of equation (2.7), we obtain the potential energy function $V = U - P\mathscr{E}$, diagonalized because of the orthogonal properties of the modes, as before. Setting $V_{11}^C = 0$ then gives the classical critical load

$$P^C = \frac{\pi^2 EI}{L^2} \tag{4.32}$$

associated with the single half-sine wave buckling mode of amplitude q_1, as described in Section 2.3.

To find the curvature of the post-buckling path (equation (4.10)) we need the coefficients V_{1111}^C and $V_{11}'^C$ determined after the elimination of passive coordinates. The second of these is given directly by the leading term of the $\mathscr{E}$ expansion, but the quartic is open to contamination from the passive coordinates, as seen later in

Chapter 8 (Figure 8.1). In this case, however, since all cubic terms are seen to vanish in the expanded forms (2.5) and (2.7), the contamination is absent, and we can write

$$V^C_{1111} = U_{1111} - P^C \mathscr{E}_{1111} = \tfrac{3}{8} EIL\left(\frac{\pi}{L}\right)^6$$

$$V'^C_{11} = -\mathscr{E}_{11} = -\frac{L}{2}\left(\frac{\pi}{L}\right)^2 \tag{4.33}$$

these being the coefficients of a Taylor expansion about C as before. The curvature of the post-buckling path is then given by

$$P^{(2)C} = -\frac{V_{1111}}{3V'_{11}}\bigg|^C = \frac{1}{4}EI\left(\frac{\pi}{L}\right)^4 \tag{4.34}$$

and we can write a first-order approximation to the path

$$P = P^C + \tfrac{1}{2}P^{(2)C}q_1^2 = EI\left(\frac{\pi}{L}\right)^2 + \frac{1}{8}EI\left(\frac{\pi}{L}\right)^4 q_1^2 \tag{4.35}$$

which is upwards-curving and stable, as expected. The analysis is extended to the next term of the series in our earlier book.[1] Other interesting studies of Euler buckling include the identification of the cusp behaviour due to Chillingworth,[75] and Zeeman's global treatment of strut and arch buckling, reprinted in his collected papers.[4]

A plate buckling model

Our second example of stable-symmetric bifurcation has a strictly bilinear load—end-shortening response, resembling the pre- and initial post-buckling of the classic plate problem. An initially stiff fundamental solution loses stability at the bifurcation point, allowing a post-buckling solution with a reduced effective stiffness on a plot of load against its corresponding deflection. The analysis illustrates first, the removal of the non-trivial fundamental path by an incremental coordinate transformation, and second the possible destabilizing influence of a non-critical (passive) coordinate.

Let us consider the system of linear springs shown in Figure 4.10, compressed by the axial load P. Buckling of the central links of length L, measured by rotation q_1, is resisted by the vertical spring of stiffness k_1, which remains vertical as the system deflects. Total end-shortening is denoted by $Q_2 L$, taken to be constant across the width corresponding to compression between rigid walls. The deflected state is entirely specified by q_1 and Q_2, which are valid generalized coordinates.

We clearly have a linear fundamental response $P = 3k_2 LQ_2$, involving pure squash of the three horizontal springs with q_1 set to zero. The post-buckling solution involves non-zero q_1, with some load being transferred from the centre to the two outside springs; this gives a second linear response, the reduced stiffness depending on the ratio k_1/k_2. We present both the complete closed-form

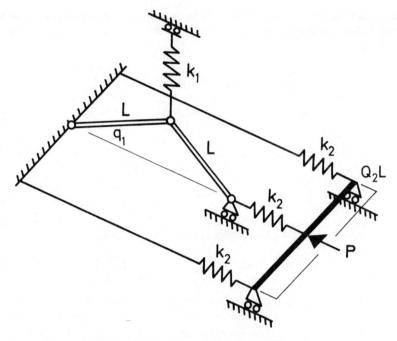

Figure 4.10 A spring and link model with a bilinear response

solution, which is only retrievable in simple cases such as this, and the more generally useful asymptotic approach.

In the general deformed configuration, the extension in the central horizontal spring is $L[Q_2 - 2(1 - \cos q_1)]$. The strain energy U is thus

$$U = \tfrac{1}{2}k_1 L^2 \sin^2 q_1 + k_2 L^2 Q_2^2 + \tfrac{1}{2}k_2 L^2 [Q_2 - 2(1 - \cos q_1)]^2 \qquad (4.36)$$

and the total potential energy becomes

$$V = U - PLQ_2$$
$$= \tfrac{1}{2}k_1 L^2 \sin^2 q_1 + k_2 L^2 Q_2^2 + \tfrac{1}{2}k_2 L^2 [Q_2 - 2(1 - \cos q_1)]^2 - PLQ_2 \qquad (4.37)$$

A sliding incremental coordinate q_2 is measured from the pure-squash fundamental solution as follows;

$$Q_2 = Q_2^F + q_2 = \frac{P}{3k_2 L} + q_2 \qquad (4.38)$$

Substituting into V, this gives

$$V = \tfrac{1}{2}k_1 L^2 \sin^2 q_1 + \tfrac{3}{2}k_2 L^2 q_2^2 - 2k_2 L^2 q_2(1 - \cos q_1)$$
$$+ 2k_2 L^2 (1 - \cos q_1)^2 - \tfrac{2}{3}PL(1 - \cos q_1) \qquad (4.39)$$

Here we have ignored terms which later vanish on differentiation with respect to q_1 and q_2.

The equilibrium equation $\partial V/\partial q_2 = 0$ can be used to eliminate q_2 as a passive coordinate. Thus

$$\frac{\partial V}{\partial q_2} = 3k_2 L^2 q_2 - 2k_2 L^2 (1 - \cos q_1) = 0 \tag{4.40}$$

which allows us to write q_2 as a function of q_1,

$$q_2 = \tfrac{2}{3}(1 - \cos q_1) \tag{4.41}$$

This can be substituted into equation (4.39) to generate a new potential function with just the single-degree-of-freedom q_1. Alternatively, we can solve both equilibrium equations simultaneously, by substituting equation (4.41) into $\partial V/\partial q_1 = 0$, to give

$$\frac{\partial V}{\partial q_1} = k_1 L^2 \sin q_1 \cos q_1 - 2k_2 L^2 q_2 \sin q_1 + 4k_2 L^2 \sin q_1 (1 - \cos q_1) - \tfrac{2}{3}PL\sin q_1$$

$$= k_1 L^2 \sin q_1 \cos q_1 + \tfrac{8}{3}k_2 L^2 \sin q_1 (1 - \cos q_1) - \tfrac{2}{3}PL\sin q_1 = 0 \tag{4.42}$$

which has two solutions

$$\sin q_1 = 0, \text{ and } P = L[4k_2 + (\tfrac{3}{2}k_1 - 4k_2)\cos q_1] \tag{4.43}$$

representing the fundamental and post-buckling equilibrium paths, respectively. Their intersection defines the critical point, given by $q_1 = 0$, $P^C = 3k_1 L/2$. We note that, by back-substitution from equations (4.41) and (4.38), the second solution may alternatively be written

$$P\left(1 - \frac{1}{4}\frac{k_1}{k_2}\right) = \tfrac{1}{2}k_1 L + L(2k_2 - \tfrac{3}{4}k_1)Q_2 \tag{4.44}$$

a linear relation between load and its corresponding deflection, the slope of which defines the effective post-buckling stiffness,

$$\frac{dP}{dQ_2} = k_2 L \frac{8k_2 - 3k_1}{4k_2 - k_1} \tag{4.45}$$

Changing the stiffnesses k_1 and k_2 provides considerable scope for varying this slope, through both positive and negative values. This is described in detail later, when the model is used to illustrate unstable-symmetric bifurcation.

Plots of P against q_1 and P against Q_2 are shown in the left and middle diagrams of Figure 4.11 for $k_1 = k_2$. These are the typical forms for a stable-symmetric point of bifurcation. The right-hand diagram shows P plotted against deflection in the vertical spring, an alternative candidate for active generalized coordinate.

It is rare that a problem can be solved in closed form, as in this simple case. More often, the potential function can only be obtained as a power series, as in the earlier strut formulation. To illustrate the associated asymptotic approach, we thus expand the sine and cosine functions of equations (4.39) as power series to give the following form:

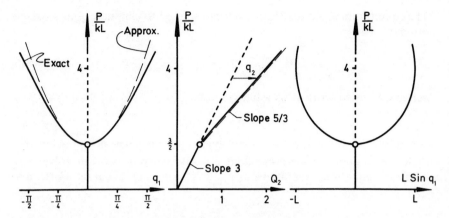

Figure 4.11 Load against active coordinate, load against end-shortening, and load against vertical deflection, for the spring model with $k = k_1 = k_2$

$$V = \tfrac{1}{2}k_1 L^2 q_1^2 + \tfrac{1}{6}(3k_2 - k_1)L^2 q_1^4 + \tfrac{3}{2}k_2 L^2 q_2^2 - k_2 L^2 q_1^2 q_2$$
$$- \tfrac{1}{3}PL q_1^2 + \tfrac{1}{36}PL q_1^4 + \text{higher-order terms} \qquad (4.46)$$

The critical point can be found directly by setting the appropriate stability coefficient $\partial^2 V / dq_1^2$ to zero:

$$V_{11}^C \equiv \left.\frac{\partial^2 V}{\partial q_1^2}\right|^C = k_1 L^2 - \tfrac{2}{3}P^C L = 0 \qquad (4.47)$$

giving $P^C = 3k_1 L/2$, as before. The following derivatives are required by a first-order post-buckling analysis;

$$V_{1111}^C \equiv \left.\frac{\partial^4 V}{\partial q_1^4}\right|^C = 4(3k_2 - k_1)L^2 + \tfrac{3}{2}P^C L = 3(4k_2 - k_1)L^2$$

$$V_{112}^C \equiv \left.\frac{\partial^3 V}{\partial q_1^2 \partial q_2}\right|^C = 2k_2 L^2, \quad V_{22}^C \equiv \left.\frac{\partial^2 V}{\partial q_2^2}\right|^C = 3k_2 L^2 \qquad (4.48)$$

$$V_{11}'^C \equiv \left.\frac{\partial^3 V}{\partial q_1^2 \partial P}\right|^C = -\tfrac{2}{3}L$$

Eliminating q_2 as a passive coordinate, we obtain derivatives of a new single-degree-of-freedom potential function $\mathscr{A}(q_1, P)$, in terms of known derivatives. Direct use of the standard results for a diagonalized potential function,[1] summarized later in Figure 8.1, gives

$$\mathscr{A}_{1111}^C = V_{1111}^C - 3\frac{(V_{112}^C)^2}{V_{22}^C} = (8k_2 - 3k_1)L^2$$

$$\mathscr{A}_{11}'^C = V_{11}'^C = -\tfrac{2}{3}L \qquad (4.49)$$

The curvature of the post-buckling path at C is thus given, from expression (4.30), by

$$P^{(2)C} = -\frac{\mathscr{A}^C_{1111}}{3.\mathscr{A}'^C_{11}} = \frac{L}{2}(8k_2 - 3k_1) \qquad (4.50)$$

The first-order post-buckling solution, obtained by using this curvature term alone in a Taylor expansion of the path, is indicated in the left-hand plot of Figure 4.11.

We return to this model later, in the discussion of unstable-symmetric bifurcation. We note that the destabilizing effect of the passive coordinate q_2 is here neatly demonstrated by the fact that $\mathscr{A}^C_{1111} < V^C_{1111}$.

4.7 UNSTABLE-SYMMETRIC POINT OF BIFURCATION

The unstable-symmetric point of bifurcation, shown earlier in Figures 3.6 and 3.8, is generated by the same potential function as the stable-symmetric point, only now V^C_{1111} is negative. We are thus slicing the dual-cusp catastrophe, exactly as shown in Figure 4.9. The secondary equilibrium path now has a negative curvature, given by the same expression (4.30) as its stable counterpart, and is unstable.

The cusp in control space is now of considerable significance, as we have seen earlier. Using the perturbation scheme of Chapter 7, or some similar technique,[1] we can write down the leading two-thirds power law term of the asymptotic equation of imperfection-sensitivity;

$$\Lambda = \Lambda^C - \tfrac{1}{2}(V^C_{1111})^{1/3}\frac{(3\dot{V}^C_1\varepsilon)^{2/3}}{V'^C_{11}} \qquad (4.51)$$

We note again that this expression is only valid if V is written in terms of a 'sliding' incremental generalized coordinate q_1, the passive coordinates having been eliminated, and that higher-order terms can be obtained by extending the perturbation analysis, as shown in our book.[1] The same result can also be obtained by truncation of equation (4.29), as before.

A shell buckling model

The complete range of behaviour for the bilinear spring model of Figure 4.10, over the full range of deflection, can be summarized in the $P - Q_2$ plot of Figure 4.12. With increasing k_1, the critical load $P^C = 3k_1 L/2$ increases, but at the expense of a progressively more unstable post-buckling response, $3k_1 = 8k_2$, marking the transition between stable and unstable bifurcation; as an example, the arrows indicate the response under rigid load, when $k_1 = 3k_2$. At $k_1 = 4k_2$ the slope of the post-buckling response is vertical on this plot, so if $k_1 > 4k_2$ the system will experience the shell-like behaviour of dynamic snaps under rigid load, as discussed in Chapter 8.

Whatever the nature of the post-buckling response, we see that the deformation is limited to $q_1 = \pi$, where we find a second symmetric bifurcation. Here the rigid links have completely reversed and buckling deformation is complete, the system then following a third equilibrium path parallel to the fundamental path, but displaced from it by an interval of 4/3 on the Q_2 axis of Figure 4.12. The system will, of course, snap dynamically to this path should it experience an unstable-symmetric point of bifurcation under dead load.

In such circumstances the equations of imperfection-sensitivity are too complicated for closed-form solution, but the asymptotic result can be quickly retrieved from equation (4.51). We suppose that the vertical spring of Figure 4.10 is initially too long by an amount εL. The only change to successive forms of the potential function (4.37) and (4.39) occurs in the first term, where $k_1 L^2(\sin^2 q_1)/2$ becomes $k_1 L^2(\sin q_1 + \varepsilon)^2/2$. Next, expanding equation (4.39) in power series form as before, equation (4.46) is replaced by

$$V = \tfrac{1}{2}k_1 L^2 q_1^2 + \tfrac{1}{6}(3k_2 - k_1)L^2 q_1^4 + \tfrac{3}{2}k_2 L^2 q_2^2 - k_2 L^2 q_1^2 q_2$$
$$- \tfrac{1}{3}PLq_1^2 + \tfrac{1}{36}PLq_1^4 + k_1 L^2 \varepsilon q_1 + \text{higher-order terms} \qquad (4.52)$$

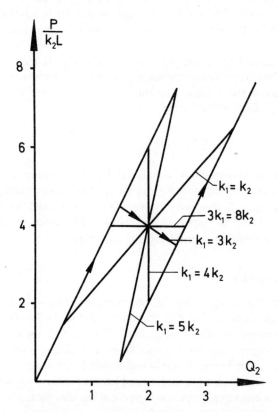

Figure 4.12 Load against end-shortening for the spring model, with varying k_1/k_2. The arrows show the response under rigid loading, for $k_1 = 3k_2$

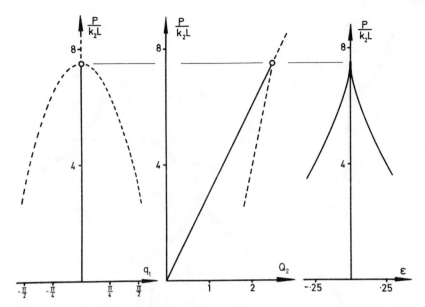

Figure 4.13 Load against active coordinate, load against end-shortening, and asymptotic form of imperfection-sensitivity, for $k_1 = 5k_2$

the only change in those terms explicitly presented being the single new contribution at the end. Expanding about the critical point C of the perfect system as before, this has the coefficient

$$\dot{V}_1^C = \left.\frac{\partial^2 V}{\partial q_1 \partial \varepsilon}\right|^C = k_1 L^2 \qquad (4.53)$$

which remains unchanged on eliminating q_2 as a passive coordinate. The remaining derivatives necessary for the first-order imperfection-sensitivity relationship are given by equation (4.49), so we can substitute directly into equation (4.51) to obtain

$$P = P^C - \tfrac{1}{2}(\mathscr{A}_{1111}^C)^{1/3}\frac{(3\dot{V}_1^C\varepsilon)^{2/3}}{\mathscr{A}_{11}^{\prime C}} = P^C - \tfrac{9}{4}L[k_1^2(k_1 - \tfrac{8}{3}k_2)]^{1/3}\varepsilon^{2/3} \qquad (4.54)$$

This is shown in Figure 4.13, together with the corresponding P–q_1 and P–Q_2 plots, for $k_1 = 5k_2$.

4.8 ROUTES THROUGH THE CUSP

As with the fold, we can see the cusp catastrophe in several different manifestations, depending on the route taken by the Λ-axis through control space. We show in Figure 4.14 typical cusped critical boundaries in the three-dimensional control space. Here the upper right-hand diagram represents the normal (stable) cusp, the lower diagram being the dual (unstable) cusp. In both cases the critical

84

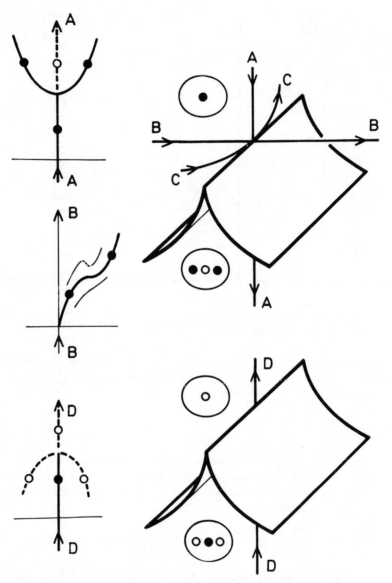

Figure 4.14 Routes in control space through the cusp, giving the stable-symmetric and unstable-symmetric points of bifurcation and the cut-off point

boundary separates the space into two regions, inside the cusp being a domain of three equilibria while outside is a domain having a single equilibrium solution.

The stability boundary of the stable cusp separates a region with a single stable state from one with one unstable and two stable states, as indicated. Route *A* then gives us the stable-symmetric point of bifurcation (upper left) while route *B* gives us a *cut-off point* (middle left); we discuss the latter in more depth in the following section.

The stability boundary of the unstable cusp separates a region with a single unstable state from one with a stable state and two unstable states, as shown. An approach similar to that of *B* is now of no physical interest, since it represents an *everywhere unstable* family of paths, and we are left with the approach *D*, which generates the unstable-symmetric point of bifurcation (lower left).

Of course, we have not described here all possible routes through these cusped surfaces. For example, we might also approach the stable cusp as in route *C* of the upper right-hand diagram. Under perturbation this would be associated with the growth from a point of a closed loop of stable states, enclosing an unstable segment of an otherwise stable path, and has been called the lip singularity in the literature. Similarly, we have the beak-to-beak singularity, where an approach like route *C* is made from inside the cusped surface. For more details of these we refer to Thom[3] and Poston and Stewart.[5]

Cut-off points and the tilted cusp

A cut-off point (or non-degenerate hysteresis point[6]) is the result of slicing the surface of Figure 4.9 across, instead of parallel to, the axis of symmetry of the cusp; it is introduced via the concept of 'routes through catastrophes' in the previous section. The cusp point itself now appears as the transition between a monotonically varying equilibrium path, and one displaying two limit points as in Figure 4.2. Cut-off points seem to appear less frequently in the responses of physical systems than the (nominally) symmetric points of bifurcation. This is perhaps because of the strong role played by symmetries, a point we shall return to in Chapter 8.

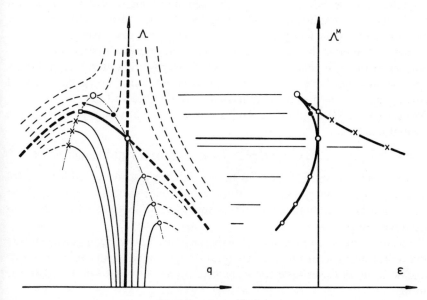

Figure 4.15 A tilted cusp

But, bifurcationally speaking, the symmetric branching points require a third control parameter for structural stability, as pointed out by Golubitsky and Schaeffer,[6] and discussed in Section 3.7. Thus if we have just two dimensions of control, we would expect a nominally symmetric bifurcation to appear as a *tilted cusp*,[76] shown here in its unstable version in Figure 4.15. Here the cusp point itself is an (unstable) cut-off point in an imperfect system, and the perfect system experiences locally an asymmetric point of bifurcation. We note that a real system would actually experience a cusp which is also off-set a general amount ξ relative to the Λ-axis; this effect is not shown here.

The careful experimental work of Roorda[53−6] on arches suggests that, while the offset of the cusp is clearly important, as seen in Figure 5.5, the magnitude of the tilt is barely discernible. We might thus conclude, as discussed in Section 4.5, that in this instance the third control acting on a cubic of the potential function is of small significance when compared with the imperfection acting on a linear term; this could be confirmed analytically by determining the imperfection-sensitivity with the third control as the imperfection. It is interesting to observe that the third control parameter contrives to break the symmetry still further, after the initial important symmetry-breaking effect of the first imperfection.

Cut-off points arise in the responses of clamped arches and domes, which exhibit limiting behaviour when deep but a monotonic response when shallow. We note that the tilted cusp is an integral part of Zeeman's model in developmental biology,[4] concerning the growth of primary and secondary waves in the differentiation of tissue type.

4.9 HIGHER-ORDER UNI-MODAL SINGULARITIES

We finally examine, fairly briefly, the two remaining higher-order cuspoids on Thom's list, the swallowtail and butterfly. As the phenomena get more complicated we could clearly devote an increasing amount of space to them, but ultimately this must be tempered by their importance in practical terms. Instances of these singularities do arise in the engineering literature, although sometimes in somewhat contrived circumstances, but overall we feel them to be of less real significance than the manifestations of the fold and the cusp.

We start with the swallowtail. This arises with the Taylor expansion of the potential function

$$V = \tfrac{1}{120} V^C_{11111} q_1^5 + \tfrac{1}{6} V^{2C}_{111} \eta q_1^3 + \tfrac{1}{2} V'^{C}_{11} \lambda q_1^2 + \dot{V}^C_1 \varepsilon q_1 + \text{higher-order terms}$$

and we see that λ and ε appear as before, but we also have a third control parameter η acting on a cubic term; this is required for structural stability of the general (non-bifurcational) form. Here a superscript 2 denotes partial differentiation with respect to η.

We now see the developing pattern of the cuspoids. Each is distinguished by the order of its leading term, and for structural stability control parameters must be introduced on all lower-order terms, except a term of order one less than the leading.

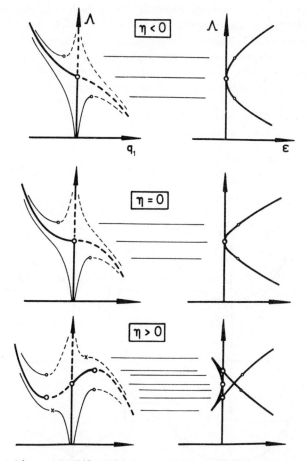

Figure 4.16 Bifurcational view of the swallowtail catastrophe with $V^C_{11111} < 0$ and $V^{2C}_{111} > 0$

Returning to the swallowtail, the coordinate-control space is now necessarily four-dimensional, and we are obliged to take sections through it. We show in Figure 4.16 a bifurcational description of the swallowtail, drawn at different but constant values of η; three-dimensional views of the corresponding equilibrium surfaces could also be sketched but are not given explicitly here. We note the appearance of cut-off points in the response of *particular* imperfect systems, shown as crosses in the bottom diagram.

The right-hand diagrams show sections through the three-dimensional control space. We have a change in failure locus, from the smooth curve of the top diagram to the intersecting form with a pair of cusps (the cut-off points) at the bottom. Drawings of the associated three-dimensional failure surface appear in the literature,[3-5] and are not reproduced here.

The swallowtail catastrophe, because of its fundamental lack of symmetry, is of less practical significance than the butterfly. However, it does arise in Zeeman's study of ship stability,[4] and a strut on a non-linear foundation can be obliged to exhibit the response of Figure 4.16 by juggling with the foundation constants.[77] But it will perhaps be known best for its appearance in the sub-strata of important higher-order singularities, such as the parabolic umbilic and the double cusp.

The butterfly catastrophe arises from the potential function

$$V = \tfrac{1}{720} V^C_{111111} q_1^6 + \tfrac{1}{24} V^{3C}_{1111} \mu q_1^4 + \tfrac{1}{6} V^{2C}_{111} \eta q_1^3 + \tfrac{1}{2} V'^C_{11} \lambda q_1^2$$
$$+ \dot{V}^C_1 \varepsilon q_1 + \text{higher-order terms}$$

where a superscript 3 denotes partial differentiation with respect to the fourth control parameter μ, acting on a quartic term, that now is required for general structural stability.

The coordinate-control space is now five-dimensional, and some repre-

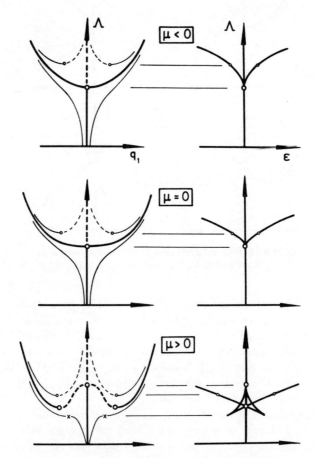

Figure 4.17 Bifurcational view of the butterfly catastrophe with $\eta = 0$, $V^C_{111111} > 0$, and $V^{3C}_{1111} < 0$

sentative bifurcational sections are shown in Figure 4.17. These are drawn for $\eta = 0$, and demonstrate the effect of varying the symmetry-preserving control μ, changing the bifurcation itself from stable-symmetric to locally unstable-symmetric but retaining the upwards-curving form in the large. Again we note the presence of cut-off points in particular imperfect systems. The effect of varying η is not demonstrated, but it would merely tilt the bifurcation in a way similar to that discussed earlier for the cusp. The right-hand diagrams illustrate the corresponding two-dimensional sections through the four-dimensional failure surface, and we see the change from a cusp, to a failure locus with three cusps and three intersections.

The butterfly catastrophe does arise in several different instances in the literature, some, as for the cusp, associated with a built-in symmetry of the system. Poston and Stewart find one in the buckling of a finite-element, axially compressible spring.[5] Similarly, Hui and Hansen generate a butterfly in the response of a strut on a non-linear elastic foundation by manipulating the spring constants.[77] It is used to describe the evolution of hull shape in Zeeman's study of ship stability.[4] Zeeman also gives the butterfly considerable prominence in his social science applications, particularly in the study of the nervous disorder, anorexia nervosa, where it is used to describe the growth of a state of normal behaviour, encouraged by trance therapy, which lies between alternating extreme states of fasting and gorging.[4]

The reduced Euler load

The butterfly catastrophe is also found in the sub-strata of higher-order phenomena, and we now briefly discuss one instance of this where, as part of an underlying but obscured double-cusp, it can play an important role in interactive buckling. It can arise in the response of a general stiffened structure, for example, as we see in Chapter 8.

We suppose that we have a system with two degrees of freedom, and a stable fundamental path which first loses its stability at a symmetric point of bifurcation C, critical with respect to mode u_1. We further suppose that a second point of bifurcation, critical with respect to u_2, appears on the fundamental path in its unstable region, close to C. For convenience the modes are assumed to be orthogonal, so the potential function is diagonalized, and the fundamental path is given by $u_1 = u_2 = 0$.

Analytically there are two alternatives. We can either analyse this as a perturbation, generated by *splitting parameter σ*, of the fully-coincident case (see Chapter 6), or we can treat the first bifurcation as a distinct, one-degree-of-freedom phenomenon and take account of the interaction via the elimination of passive coordinates. Clearly, the former is the more complete, and gives the better global picture, but the latter can also yield important local information, and it is here that we find the butterfly.

The elimination of passive coordinates gives the two derivatives necessary to evaluate the curvature of the post-buckling path at C:

$$\mathscr{A}'^C_{11} = A'^C_{11}, \quad \mathscr{A}^C_{1111} = A_{1111} - \frac{3(A_{112})^2}{A_{22}}\Bigg|^C$$

Here A is the original two-degree-of-freedom diagonalized potential function and $\mathscr{A}$ is the equivalent one-degree-of-freedom function defined by eliminating u_2 as a passive coordinate, according to the process of Chapter 7, summarized in Figure 8.1. Whether C is a stable or an unstable-symmetric point of bifurcation depends, of course, on the sign of $\mathscr{A}^C_{1111}$, as we have seen earlier in equation (4.30).

The implications should now be clear. C is stable with respect to u_2, so A^C_{22} must be positive, and taking the cross-term A^C_{112} to be non-zero we have $\mathscr{A}^C_{1111} < A^C_{1111}$. We reach the important conclusion that the second bifurcation can only have a destablizing effect on the first. Moreover, as the critical points approach each other, A^C_{22} approaches zero, and therefore the destabilizing effect will grow. Thus, at some finite separation, it counteracts the effect of a positive A^C_{1111} and we have a point of zero path curvature—the butterfly point itself. Clearly, the separation parameter σ of the two bifurcations plays the same role here as the control parameter μ of Figure 4.17, although we would not expect it to act directly on a fourth-order term as does μ. This is discussed further in Section 8.5.

A closely associated phenomenon can be found in the buckling of a simple Shanley column, developed specifically to model certain stiffened plates.[78] Here a bilinear spring is used to allow for two differing responses, a pure overall (Euler) mode and a mode combining overall and local buckling. When the plate is in its combined mode, the response is related to a reduced-modulus column, which either buckles in stable or unstable fashion, depending on the ratio of local to overall buckling load. At a certain ratio the buckling is effectively neutral, arising at what is termed the *reduced Euler load*. The neutral nature of the instability is entirely a result of the limitations of the model, no allowance being made for the equivalent of sixth-order terms in the analysis. It provides, in fact, a localized view of the more general butterfly discussed above.

5

Buckling and Imperfection-Sensitivity of Arches

5.1 SIMPLIFICATION VIA INEXTENSIBILITY

The cusp geometry dictates the response of many elastic structures, largely because of inherent symmetries in the associated perfect systems. To illustrate this we now develop the potential function for a simplified elastic arch model under central loading, and analyse it in the light of the previous chapter. The approach is considerably influenced by the impressive topological study of associated strut and arch buckling of Zeeman.[4] It is pleasant for us, as engineers, to be able to acknowledge the extra insight that we have gained from this view through the eyes of an eminent mathematician.

The response of the structure is clearly non-linear, but, as in the strut formulation of Chapter 2, we assume that a small element of the arch in pure bending obeys Hooke's law; this gives the strain energy for a general deformed shape. The arch is assumed inextensional, thereby eliminating awkward compressibility effects which contribute nothing phenomenologically but clutter up the final results.[44]

After the formulation in general terms a Fourier series expansion of the deformed shape is assumed. This differs from the earlier strut analysis in that just the leading two terms are included. We find that the assumption greatly simplifies the *pre-buckling* analysis, and is later justified by good agreement with experiment. The inextensibility, coupled with a constraint condition imposed by the fixed supports, then reduces the degrees of freedom of the mathematical model to one.

Finally, an extra dimension is introduced by allowing for the effect of pre-stress. We oblige the arch to take up a particular deformed shape, and then assume that it is somehow stress-relieved. Thus we can compare the arch which is stress-free under no load, with an initially straight strut buckled into an arch between fixed supports, or any other of a wide range of pre-stress conditions.

5.2 STRAIN ENERGY WITH ARBITRARY PRE-STRESS

For the strain energy we proceed in the same way as for the inextensional strut of Chapter 2, but allowing for the extra effect of full stress-relief in a general

92

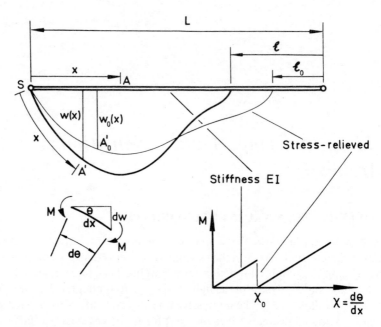

Figure 5.1 General deformation of the inextensional arch, and the stress-relieved state

deformed state. Let us consider the initially straight, simply-supported, strut of length L shown in Figure 5.1, and oblige it to undergo an end-displacement $\mathscr{E}$ in the axial direction as shown. We assume that the strut is inextensional, and has a flexural stiffness EI.

The curvature of a small element, as in Chapter 2, is given by

$$\chi = \frac{\mathrm{d}\theta}{\mathrm{d}x} = \frac{\mathrm{d}}{\mathrm{d}x}\sin^{-1}w' = w''(1 - w'^2)^{-1/2}$$

where a prime denotes differentiation with respect to x. We have only strain energy of bending to consider, which is given by

$$\mathrm{d}U = \tfrac{1}{2}M\mathrm{d}\theta = \tfrac{1}{2}EI\chi\mathrm{d}\theta = \tfrac{1}{2}EI\chi^2\mathrm{d}x$$

We now suppose that, at a certain deformed shape $w_0(x)$ with end-displacement $\mathscr{E}_0$ as shown, the system is *stress-relieved*, such that the elemental bending moment M and the stored strain energy drop to zero. The moment–curvature relationship takes the form shown in Figure 5.1, and we can see that the elemental strain energy for a general deformed shape $w(x)$ is now

$$\mathrm{d}U = \tfrac{1}{2}EI(\chi - \chi_0)^2\mathrm{d}x$$

where χ_0 is the curvature in the stress-relieved state. Integrating, we can now write the total strain energy as

$$U = \tfrac{1}{2}EI \int_0^L (\chi - \chi_0)^2 \, dx$$

$$\tfrac{1}{2}EI \int_0^L [w''^2(1 - w'^2)^{-1} - 2w'' w_0''(1 - w'^2)^{-1/2}(1 - w_0'^2)^{-1/2}$$

$$+ w_0''^2(1 - w_0'^2)^{-1}] \, dx$$

Expanding all these terms as power series, we have

$$U = \tfrac{1}{2}EI \int_0^L [(w_0''^2 + w_0''^2 w_0'^2 + w_0''^2 w_0'^4 + \cdots)$$

$$- 2w_0''(1 + \tfrac{1}{2}w_0'^2 + \tfrac{3}{8}w_0'^4 + \cdots)w'' + w''^2$$

$$- w_0''(1 + \tfrac{1}{2}w_0'^2 + \tfrac{3}{8}w_0'^4 + \cdots)w'' w'^2 + w''^2 w'^2 + \cdots] \, dx \qquad (5.1)$$

which is arranged in ascending powers of w and its derivatives. For a linear eigenvalue study we would need merely the quadratic terms.

The corresponding total end constraint $\mathscr{E}$ is given by

$$\mathscr{E} = L - \int_0^L \cos\theta \, dx$$

$$= \int_0^L (\tfrac{1}{2}w'^2 + \tfrac{1}{8}w'^4 + \tfrac{1}{16}w'^6 + \cdots) \, dx \qquad (5.2)$$

as in Chapter 2. Again, a linear eigenvalue study needs just the leading term.

5.3 EXPANSION IN FOURIER HARMONICS

After the general formulation we now restrict the deflection functions $w_0(x)$ and $w(x)$ to appropriate Fourier harmonics. We suppose that the arch is initially deflected upwards with a rise H at the crown, and is loaded perfectly centrally by a dead vertical load P. Later we test for the effects of imperfections by allowing a small offset of the load, but for the moment we restrict the study to this *perfect system*.

We start by assuming that in the stress-relieved state, before the arch is fixed between the rigid supports, the deflection $w_0(x)$ is a half-sine wave, as shown at the top of Figure 5.2. We thus write

$$w_0 = A_0 \sin\frac{\pi x}{L}$$

where A_0, like w_0, is positive measured downwards. This allows for the study of a variety of states of initial pre-stress by altering the value of A_0. We note in passing that a symmetry-destroying imperfection could be introduced here, in the form of pre-stress, with the addition of a small second-harmonic term.

94

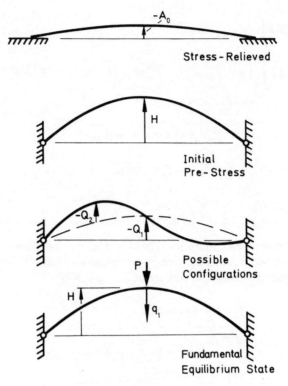

Figure 5.2 Arch geometry and description

For a full harmonic analysis of the arch we could expand w as

$$w(x) = \sum_{i=1}^{\infty} Q_i \sin \frac{i\pi x}{L}$$

However, we truncate the series after the leading two terms, and write

$$w(x) = Q_1 \sin \frac{\pi x}{L} + Q_2 \sin \frac{2\pi x}{L} \tag{5.3}$$

where Q_1 and Q_2, like w, are positive measured downwards. This reduces the system to one which at first seems to have two degrees of freedom. But the inextensibility and the constraint of the fixed supports impose a further condition so that Q_1 and Q_2 are not independent; we thus have some locus in Q_i space, $Q_1 = f(Q_2)$, say, which the system is obliged to follow, and it is reduced to one with a single degree of freedom.

This may seem a somewhat crude approximation to a continuum with an infinite number of degrees of freedom, but it is remarkably successful, as indicated first, by the simple closed-form solutions obtained and second, by the good agreement with experiments. It clearly closely resembles a single-degree-of-freedom Rayleigh–Ritz approach. Its great virtue is its simplicity, especially in that it exhibits a trivial fundamental path of equilibria, $Q_2 = 0$, $Q_1 = -H =$ constant, for all loads P. This is a significant aid to solution, since the description of the nonlinear path for the true arch is in itself a considerable problem, and one which is largely irrelevant as far as the bifurcational response is concerned. These are important points, and we shall be returning to them at intervals later.

So, substituting the assumed forms for w_0 and w into the general strain energy function (5.1) we are left with a number of integrals of sinusoidal products to evaluate. These exhibit both quadratic and higher-order orthogonalities, and lead to the diagonalized strain energy function

$$U(Q_1, Q_2) = \tfrac{1}{2}EI\left\{ -\frac{\pi^4}{L^2}\left[\frac{A_0}{L} + \frac{\pi^2}{8}\left(\frac{A_0}{L}\right)^3 + \text{higher-order} \right]Q_1 \right.$$
$$+ \frac{1}{2}\frac{\pi^4}{L^3}(Q_1^2 + 16Q_2^2) - \frac{1}{8}\frac{\pi^6}{L^4}\left(\frac{A_0}{L} + \text{higher-order} \right)(Q_1^3 + 8Q_1 Q_2^2)$$
$$\left. + \frac{1}{8}\frac{\pi^6}{L^5}(Q_1^4 + 40Q_1^2 Q_2^2 + 64Q_2^4) + \text{higher-order} \right\} \qquad (5.4)$$

which is arranged in ascending powers of Q_1 and Q_2. Here constant terms arising from just w_0 and its derivatives have been ignored since we are concerned with variations of the strain energy, never its absolute value.

5.4 THE CONSTRAINT CONDITION

We now turn our attention to the constraint condition imposed by the rigid supports, and seek a way of expressing the resulting locus in Q_1–Q_2 space. This is to be written as a Taylor series, first in the two variables, and second via an intrinsic perturbation scheme, in Q_2 alone. The latter is only appropriate if the locus is single-valued in Q_2, although during buckling Q_2 reaches a maximum and then falls again to zero as the arch finds its final, upsidedown, equilibrium configuration. The analysis is thus essentially localized, which falls into line with the philosophy of structural mechanics, where interest focuses more on the failure itself than on the gross deformations of a failed system. The treatment can be seen as a useful quantitative complement to Zeeman's topological study.[4]

The intrinsic perturbation scheme[79,80] is a concept we shall use at length in the general branching analysis of Chapter 7, and this simple example serves as an introduction to the more advanced studies. Only two variables are involved, and we can write all derivatives explicitly without having to resort to the neater, but more obscure, subscript notation used elsewhere in the book.

The constraint condition is simply $\mathscr{E} = \text{constant}$, where $\mathscr{E}$ is the end-shortening function (5.2). Substituting the assumed form for w, given by equation (5.3), into this expression and performing the necessary integrations, we have

$$\mathscr{E} = \frac{\pi^2}{4L}(Q_1^2 + 4Q_2^2) + \frac{3}{64}\frac{\pi^4}{L^3}(Q_1^4 + 16Q_1^2 Q_2^2 + 16Q_2^4) + \text{higher-order terms} \qquad (5.5)$$

We see that $\mathscr{E}$ is diagonalized (no $Q_1 Q_2$ cross-term), and higher-order orthogonalities have set some of the quartics to zero.

We are seeking a Taylor expansion, and thus must start with coordinates measured from the unbuckled configuration. Introducing the incremental coordinate q_1 defined by

$$Q_1 = -H + q_1 \tag{5.6}$$

(see Figure 5.2), we substitute this into equation (5.5) to give

$$\mathscr{E} = \frac{1}{4}\frac{\pi^2}{L}(q_1^2 - 2Hq_1 + 4Q_2^2) + \frac{3}{64}\frac{\pi^4}{L^3}(q_1^4 - 4Hq_1^3 + 6H^2q_1^2 - 4H^3q_1$$

$$+ 16q_1^2 Q_2^2 - 32Hq_1 Q_2^2 + 16H^2 Q_2^2 + 16Q_2^4) + \text{higher-order terms} \tag{5.7}$$

Here as before, constant terms are ignored, since we are to be concerned just with variations of $\mathscr{E}$, never its absolute value. We see that the expression has become more complicated, with the appearance of linear and cubic terms.

But we have now a Taylor series, and can immediately write down the derivatives of $\mathscr{E}$, evaluated in the pre-buckled state $F (q_1 = Q_2 = 0)$, that we need later. We have, for example,

$$\left.\frac{\partial \mathscr{E}}{\partial q_1}\right|^F = -\frac{1}{2}\frac{\pi^2 H}{L}\left[1 + \text{order}\left(\frac{H}{L}\right)^2\right], \quad \left.\frac{\partial \mathscr{E}}{\partial Q_2}\right|^F = 0$$

$$\left.\frac{\partial^2 \mathscr{E}}{\partial q_1^2}\right|^F = \frac{1}{2}\frac{\pi^2}{L}\left[1 + \text{order}\left(\frac{H}{L}\right)^2\right], \quad \left.\frac{\partial^2 \mathscr{E}}{\partial Q_2^2}\right|^F = 2\frac{\pi^2}{L}\left[1 + \text{order}\left(\frac{H}{L}\right)^2\right] \text{etc}$$

and we need the cubic

$$\left.\frac{\partial^3 \mathscr{E}}{\partial q_1 \partial Q_2^2}\right|^F = -3\frac{\pi^4 H}{L^3}\left[1 + \text{order}\left(\frac{H}{L}\right)^2\right]$$

and the quartic

$$\left.\frac{\partial^4 \mathscr{E}}{\partial Q_2^4}\right|^F = 18\frac{\pi^4}{L^3}\left[1 + \text{order}\left(\frac{H}{L}\right)^2\right]$$

In all the following analysis only the leading term of each series need be considered, since we assume the arch to be of moderate rise, and hence

$$\left(\frac{H}{L}\right)^2 \ll 1$$

A simple perturbation scheme

We start the perturbation analysis by assuming that the system is constrained to follow a locus in coordinate space described by the parametric form

$$q_1 = q_1(Q_2)$$

This is substituted into the constraint condition $\mathscr{E} = \text{constant} = K$, which then becomes an identity, since it is satisfied by all values of the single independent variable Q_2 in the region of interest. Thus

$$\mathscr{E}[q_1(Q_2), Q_2] \equiv K \tag{5.8}$$

This can be repeatedly differentiated with respect to Q_2. Evaluation of the resulting equations in the unbuckled state F then gives a series of sequentially linear problems, which are successively solved for derivatives of q_1 with respect to Q_2. These could be used to construct a Taylor series form of the constraint condition in the one variable Q_2, or we can use the derivatives directly, as in the following post-buckling study.

We note that the process can only be applied to an identity, or we would be evaluating (on the locus of interest) before differentiating, thereby denying the system its full range of allowable deflection configurations.

Thus differentiating equation (5.8) repeatedly, we obtain

$$\frac{\mathrm{d}\mathscr{E}}{\mathrm{d}Q_2} = \frac{\partial\mathscr{E}}{\partial q_1}\frac{\mathrm{d}q_1}{\mathrm{d}Q_2} + \frac{\partial\mathscr{E}}{\partial Q_2} = 0$$

$$\frac{\mathrm{d}^2\mathscr{E}}{\mathrm{d}Q_2^2} = \frac{\partial^2\mathscr{E}}{\partial q_1^2}\left(\frac{\mathrm{d}q_1}{\mathrm{d}Q_2}\right)^2 + 2\frac{\partial^2\mathscr{E}}{\partial q_1 \partial Q_2}\frac{\mathrm{d}q_1}{\mathrm{d}Q_2} + \frac{\partial^2\mathscr{E}}{\partial Q_2^2} + \frac{\partial\mathscr{E}}{\partial q_1}\frac{\mathrm{d}^2q_1}{\mathrm{d}Q_2^2} = 0 \text{ etc.}$$

In the following work we require the first four equations, but the last two are lengthy and are not given explicitly here.

Evaluating at F and using the known partial derivatives of $\mathscr{E}$, the first three equations give directly

$$\left.\frac{\mathrm{d}q_1}{\mathrm{d}Q_2}\right|^F = 0, \quad \left.\frac{\mathrm{d}^2q_1}{\mathrm{d}Q_2^2}\right|^F = \frac{4}{H}, \quad \left.\frac{\mathrm{d}^3q_1}{\mathrm{d}Q_2^3}\right|^F = 0 \tag{5.9}$$

and the fourth equation is reduced to

$$\left.\frac{\partial^4\mathscr{E}}{\partial Q_2^4} + 6\frac{\partial^3\mathscr{E}}{\partial q_1 \partial Q_2^2}\frac{\mathrm{d}^2q_1}{\mathrm{d}Q_2^2} + 3\frac{\partial^2\mathscr{E}}{\partial q_1^2}\left(\frac{\mathrm{d}^2q_1}{\mathrm{d}Q_2^2}\right)^2 + \frac{\partial\mathscr{E}}{\partial q_1}\frac{\mathrm{d}^4q_1}{\mathrm{d}Q_2^4}\right|^F = 0 \tag{5.10}$$

Substituting the known derivatives, the first two terms can be neglected according to the moderate rise approximation, $(H/L)^2 \ll 1$. This leads to the simple result for the fourth derivative,

$$\left.\frac{\mathrm{d}^4q_1}{\mathrm{d}Q_2^4}\right|^F = \frac{48}{H^3} \tag{5.11}$$

It is interesting to observe that this could have been achieved using no higher than quadratic derivatives of $\mathscr{E}$. Here we abandon the scheme, although further derivatives can be found by the obvious extension, if necessary.

5.5 LINEAR EIGENVALUE ANALYSIS

Before we proceed with the eigenvalue analysis to determine the critical load of the arch we first write down the full potential energy expression, including the non-linear terms required later in the post-buckling analysis. Combining the strain energy (equation (5.4)) and the work done by the load $(-Pq_1)$, and adopting the earlier incremental transformation (equation (5.6)) so we have a Taylor series as before, the potential energy function can be written

$$V = \tfrac{1}{2}EI\left\{ -\frac{\pi^4}{L^2}\left[\frac{A_0}{L} + \frac{H}{L} + \frac{\pi^2}{8}\left(\frac{A_0}{L}\right)^3 + \tfrac{3}{8}\pi^2\frac{A_0}{L}\left(\frac{H}{L}\right)^2 + \frac{\pi^2}{2}\left(\frac{H}{L}\right)^3\right]q_1 \right.$$

$$+ \frac{1}{2}\frac{\pi^4}{L^3}(q_1^2 + 16Q_2^2) - \frac{1}{8}\frac{\pi^6}{L^4}\left(\frac{A_0}{L} + 4\frac{H}{L}\right)q_1^3 - \frac{\pi^6}{L^4}\left(\frac{A_0}{L} + 10\frac{H}{L}\right)q_1Q_2^2$$

$$\left. + \frac{1}{8}\frac{\pi^6}{L^5}(q_1^4 + 40q_1^2Q_2^2 + 64Q_2^4) + \text{higher-order terms}\right\} - Pq_1 \qquad (5.12)$$

Here, as before, constant terms have been ignored, and we have assumed that the arch in both the stress-relieved and pre-stressed states is of moderate rise, thereby neglecting $(A_0/L)^2$, $(H/L)^2$, and A_0H/L^2 in comparison with unity. However, higher-order terms have been included in the coefficient of the linear q_1 term, since we shall later investigate specifically an arch with no initial pre-stress, for which the leading term vanishes; this turns out to be an unnecessary precaution, but for the moment we retain the terms.

For any constant load P this is a Taylor series in q_1 and Q_2, and we can immediately write down all the partial derivatives that we shall need later. We have first,

$$\left.\frac{\partial V}{\partial q_1}\right|^F = -\tfrac{1}{2}EI\frac{\pi^4}{L^2}\left\{\frac{A_0}{L} + \frac{H}{L} + \frac{\pi^2}{8}\left(\frac{A_0}{L}\right)^3 + \tfrac{3}{8}\pi^2\frac{A_0}{L}\left(\frac{H}{L}\right)^2 + \frac{\pi^2}{2}\left(\frac{H}{L}\right)^3\right\} - P$$

$$\left.\frac{\partial V}{\partial Q_2}\right|^F = 0, \quad \left.\frac{\partial^2 V}{\partial q_1^2}\right|^F = \tfrac{1}{2}EI\frac{\pi^4}{L^3}, \quad \left.\frac{\partial^2 V}{\partial q_1 \partial Q_2}\right|^F = 0, \quad \left.\frac{\partial^2 V}{\partial Q_2^2}\right|^F = 8EI\frac{\pi^4}{L^3}, \text{ etc.} \qquad (5.13)$$

We shall also need the cubic coefficient

$$\left.\frac{\partial^3 V}{\partial q_1 \partial Q_2^2}\right|^F = -EI\frac{\pi^6}{L^4}\left(\frac{A_0}{L} + 10\frac{H}{L}\right) \qquad (5.14)$$

and the quartic

$$\left.\frac{\partial^4 V}{\partial Q_2^4}\right|^F = 96EI\frac{\pi^6}{L^5} \qquad (5.15)$$

in the post-buckling analysis.

It must be remembered that configurations of the arch are subject to the constraint condition, so q_1 and Q_2, as they appear here, are not valid generalized coordinates, the system having just one degree of freedom. However, remembering the earlier parametric representation of the constraint condition, $q_1 = q_1(Q_2)$, we can substitute this and regard Q_2 as the single valid variable; we note that the function is already known implicitly, as derivatives evaluated at the unbuckled state F.

Thus writing V as

$$V = V[q_1(Q_2), Q_2]$$

at any fixed load level, we differentiate with respect to Q_2 to obtain the equilibrium equation

$$\frac{dV}{dQ_2} = \frac{\partial V}{\partial q_1}\frac{dq_1}{dQ_2} + \frac{\partial V}{\partial Q_2} = 0 \tag{5.16}$$

If we evaluate this equation at F we find on substitution from equations (5.9) and (5.13) that the condition is identically satisfied for all P. This confirms that there is a trivial fundamental equilibrium state $F(Q_1 = -H, Q_2 = 0)$ for any load value.

To check the stability of this equilibrium state we form the second derivative

$$\frac{d^2V}{dQ_2^2} = \frac{\partial^2 V}{\partial q_1^2}\left(\frac{dq_1}{dQ_2}\right)^2 + 2\frac{\partial^2 V}{\partial q_1 \partial Q_2}\left(\frac{dq_1}{dQ_2}\right) + \frac{\partial^2 V}{\partial Q_2^2} + \frac{\partial V}{\partial q_1}\frac{d^2 q_1}{dQ_2^2} \tag{5.17}$$

Evaluating this at F gives the single relevant *stability coefficient*

$$\left.\frac{d^2V}{dQ_2^2}\right|^F = 8EI\frac{\pi^4}{L^3} - \left\{\frac{1}{2}EI\frac{\pi^4}{L^2}\left[\frac{A_0}{L} + \frac{H}{L} + \frac{\pi^2}{8}\left(\frac{A_0}{L}\right)^3 + \frac{3}{8}\pi^2\frac{A_0}{L}\left(\frac{H}{L}\right)^2\right.\right.$$
$$\left.\left. + \frac{\pi^2}{2}\left(\frac{H}{L}\right)^3\right] + P\right\}\frac{4}{H}$$

which we set to zero to find the *critical equilibrium state C*. This gives, after a little algebra

$$P^C = \frac{1}{2}EI\frac{\pi^4}{L^2}\left\{3\frac{H}{L} - \frac{A_0}{L} - \frac{\pi^2}{8}\left(\frac{A_0}{L}\right)^3 - \frac{3}{8}\pi^2\frac{A_0}{L}\left(\frac{H}{L}\right)^2 - \frac{\pi^2}{2}\left(\frac{H}{L}\right)^3\right\}$$

The last three, higher-order, terms can now be safely neglected, since the case of a vanishing leading term ($A_0 = 3H$) is only of passing interest. Thus

$$P^C = \frac{1}{2}\frac{\pi^4 EI}{L^3}(3H - A_0) \tag{5.18}$$

We note that for the range of pre-stress $A_0 \geq 3H$ we have $P^C \leq 0$, and can tentatively conclude that an arch in this range of heavy pre-stress in the sagging sense cannot be persuaded into a hogging configuration even under zero load. However, the transitional case $A_0 = 3H$ is precisely that at which the leading term vanishes, so this conclusion is subject to higher-order effects which we explore no further here.

•

The arch formed from an initially straight strut

Here $A_0 = 0$, since in the stress-relieved state the arch is straight. Thus we have the simple result for the critical load

$$P^C = \frac{3}{2} \frac{\pi^4 EIH}{L^3} \tag{5.19}$$

The arch with no initial pre-stress

Here the stress-relieved state coincides with the pre-buckled state and $A_0 = -H$. This gives the critical load

$$P^C = 2 \frac{\pi^4 EIH}{L^3}$$

Higher-order derivatives

It is convenient at this stage to evaluate some of the higher derivatives of V that are required in the post-buckling analysis. This is for two reasons. First, they follow naturally from the above differentiation process, but second, and more important, we thus avoid possible confusion over a significant notational change that follows later. The derivatives are to be evaluated at the critical point C.

Thus differentiating equation (5.17) once with respect to Q_2 and evaluating at C we obtain

$$\frac{d^3 V}{dQ_2^3}\bigg|^C = 0$$

This can be quickly confirmed from symmetry considerations,[81] as discussed in Section 8.2. Differentiating a second time before evaluation gives

$$\frac{d^4 V}{dQ_2^4}\bigg|^C = \frac{\partial^4 V}{\partial Q_2^4} + 6 \frac{\partial^3 V}{\partial q_1 \partial Q_2^2} \frac{d^2 q_1}{dQ_2^2} + 3 \frac{\partial^2 V}{\partial q_1^2} \left(\frac{d^2 q_1}{dQ_2^2}\right)^2 + \frac{\partial V}{\partial q_1} \frac{d^4 q_1}{dQ_2^4}\bigg|^C \tag{5.21}$$

on evaluation at C. Substituting now from the known derivatives we find, perhaps rather surprisingly, that the leading two terms of the right-hand side can be neglected according to the assumptions for moderate rise, $(H/L)^2 \ll 1$, etc. This suggests a result that is later confirmed, that no derivatives of the original potential function (written in terms of q_1 and Q_2) of order higher than quadratic are required to find the post-buckling path curvature. Thus the dominant feature of arch behaviour is the enforced geometry change which arises from the constraint condition.

We thus obtain the result for the fourth derivative

$$\frac{d^4 V}{dQ_2^4}\bigg|^C = -48 \frac{P^C}{H^3} - 24 EI \frac{\pi^4 A_0}{L^3 H^3}$$

and substitution of the general form for P^C of equation (5.18) gives

$$\left.\frac{d^4 V}{dQ_2^4}\right|^C = -72\frac{\pi^4 EI}{H^2 L^3} \tag{5.22}$$

We see that the derivative is not dependent on the pre-stress amplitude A_0.

5.6 POST-BUCKLING ANALYSIS

At this stage it is rewarding to reflect on just what we know about the response of the system. This is summarized schematically in Figure 5.3. The constraint condition gives a curved surface in $P-Q_1-Q_2$ space, on which all allowable equilibrium states must lie. By symmetry, the surface forms part of a complete cylinder which recuts the Q_1 axis at $Q_1 = +H$, but we show here the range of validity of the present analysis in which Q_2 remains single-valued. We note also that the cylinder extends indefinitely in both directions to $P = +\infty$ and $-\infty$.

The fundamental equilibrium state corresponds to a path lying on the cylinder at $Q_1 = -H$, as shown, which becomes unstable at, or just above, the critical

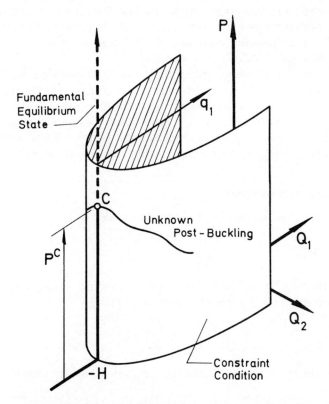

Figure 5.3 Information gained from a linear eigenvalue analysis

point C where $P = P^C$. We cannot yet say anything about the stability of the critical point itself, nor of the nature of the post-buckling equilibrium path which, by the earlier basic theorems of Section 3.3, must pass through C. These must wait for the post-buckling analysis for clarification.

But the post-buckling path must have a zero slope because of the inherent symmetry, and assuming that it has a non-zero curvature we can therefore expect either a cusp or a dual cusp catastrophe at C. We shall call on the results of Chapter 4 to determine the curvature.

Obligatory notational change

The general results of Chapter 4 are derived using a potential function which would be written here as

$$V = V(Q_2, P)$$

This differs in two significant ways from the form used in the eigenvalue analysis. First, we have no variation with respect to q_1, so we must assume that at this stage q_1 is fully replaced by its Taylor expansion in terms of Q_2, developed from the constraint condition. Second, we see that the load P must now be considered as a variable, since a derivative of V with respect to P is required in the post-buckling analysis. We can, however, make use of the differentiations of the eigenvalue analysis, but *subject to a change of notation*.

Thus, inspecting the earlier differentiation of V, we find that full derivatives with respect to Q_2, those on the left-hand side of equations (5.16) and (5.17) for example, must now be written as partial derivatives, since P was previously assumed constant. On the other hand, partial differentiation with respect to Q_2 by this stage implies varying q_1, since the constraint condition is now automatically satisfied. The meaning of the partial ∂ is thus significantly altered.

Post-buckling path curvature

With the notational change, we have the fourth derivative

$$\left.\frac{\partial^4 V}{\partial Q_2^4}\right|^C = -72\frac{\pi^4 EI}{H^2 L^3} \tag{5.23}$$

from equation (5.22), and to find the only outstanding necessary derivative we differentiate equation (5.17) with respect to P. This gives

$$\left.\frac{\partial^3 V}{\partial Q_2^2 \partial P}\right|^C = -\left.\frac{d^2 q_1}{dQ_2^2}\right|^C = -\frac{4}{H} \tag{5.24}$$

We can therefore immediately write the curvature of the post-buckling path as

$$\left.\frac{d^2 P}{dQ_2^2}\right|^C = -\left.\frac{\partial^4 V}{\partial Q_2^4}\right/3\left.\frac{\partial^3 V}{\partial Q_2^2 \partial P}\right|^C = -6\frac{\pi^4 EI}{HL^3} \tag{5.25}$$

using the general result (equation (4.30)) of Chapter 4. The curvature is negative, so we have an unstable-symmetric point of bifurcation at a dual cusp catastrophe. We note that the curvature is not dependent on the amount of initial pre-stress A_0.

5.7 THE REAL PERFECT RESPONSE

We are now in a position to complete the picture of Figure 5.3, and to compare this with the known response of the perfect arch. We consider here only the arch formed from an initially straight strut.

From symmetry, the paths clearly must be as shown on the left of Figure 5.4. We have a trivial fundamental solution $Q_1 = -H$, $Q_2 = 0$, together with its mirror image $Q_1 = H$, $Q_2 = 0$, and the constraint condition generates a cylinder extending to $P = \pm \infty$. The unstable post-buckling path forms a closed loop on the cylinder. The system, on reaching the unstable critical state C or its reflection, snaps dynamically to the opposite trivial stable state, in accordance with the constraint condition.

We contrast this picture with the true response of the shallow arch, shown on the right of Figure 5.4. The major difference is the form of the fundamental path; now, with the reintroduction of extensibility, the arch can display limiting behaviour in a symmetric mode ($Q_2 = 0$), giving a highly non-linear, non-trivial path. The limit points are, of course, already unstable with respect to Q_2. Clearly, we cannot now represent the constraint condition in any simple way. The symmetric response is discussed in more detail in Section 5.11.

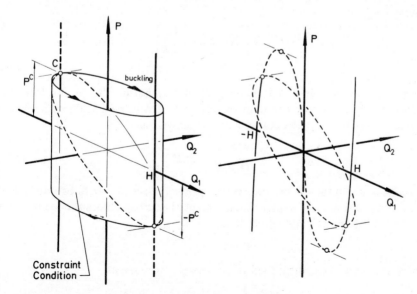

Figure 5.4 Comparison between the response of the model and the true arch, for the perfect system

But the post-buckling behaviour is essentially the same, and it is precisely the trivial nature of the fundamental path that gives such simple results. We might suppose that the further apart that the critical bifurcation and limit points are, the more accurate the analysis is likely to be. We note finally that an elastically tied arch can be adjusted, by introducing an extra control parameter, so that the two critical states coincide. This gives the hilltop branching point, which is investigated further in Chapter 6.

5.8 IMPERFECTION-SENSITIVITY ANALYSIS

The imperfection-sensitivity of the arch is first investigated following Roorda's classic formulation,[53] by off-setting the load a small amount εL from the centre-line; here we take ε as positive for off-sets in the positive sense of x. The strain energy is clearly unaffected by this introduction of a second control parameter, but the load now drops a further small amount δw, which is given by the approximation

$$\delta w \approx \varepsilon L \frac{\mathrm{d}w}{\mathrm{d}x}\bigg|_{x=L/2}$$

We can replace this by a rigorous equality, by assuming that the load is applied a distance εL along a rigid straight bar, rigidly fixed to the arch at the centre. Alternatively, we can assume that ε is small enough so that, in comparison with the earlier assumptions, such rigour is unnecessary and the equality is valid for a simple load off-set. We note that, experimentally, the former usually applies.

The potential energy of the load is changed by this additional drop, and we have

$$V = U - P\left[q_1 + \varepsilon L w'\left(\frac{L}{2}\right)\right]$$
$$= U - P(q_1 - 2\pi\varepsilon Q_2) \tag{5.26}$$

using the assumed deflected shape (5.3). Remembering the results from the constraint condition, we can immediately write down the additional necessary coefficient for the imperfection-sensitivity analysis as

$$\frac{\partial^2 V}{\partial Q_2 \partial \varepsilon}\bigg|^C = 2\pi P^C \tag{5.27}$$

where C refers to the critical state of the *perfect system*, $\varepsilon = 0$. Substituting now in the general result of Chapter 4 (equation (4.51)), we can obtain the coefficient of the two-thirds power law failure locus

$$P = P^C - \beta\varepsilon^{2/3}$$

where P is now the critical load of an *imperfect system*, as

$$\beta = \left(\frac{\partial^4 V}{\partial Q_2^4}\right)^{1/3}\left(3\frac{\partial^2 V}{\partial Q_2 \partial \varepsilon}\right)^{2/3}\bigg/2\frac{\partial^3 V}{\partial Q_2^2 \partial P}\bigg|^C$$

$$= \tfrac{3}{4}EI\frac{\pi^4}{L^3}\{3\pi^2 H(3H - A_0)^2\}^{1/3} \tag{5.28}$$

on substitution from equations (5.23), (5.24), and (5.26).

The arch formed from an initially straight strut

Here $A_0 = 0$ as before, and we thus have the two-thirds power law imperfection-sensitivity relationship

$$P = P^C = \frac{9\pi^4 EIH}{4\ L^3}(\pi\varepsilon)^{2/3}$$

Non-dimensionalizing with respect to P^C, given by equation (5.19), we have

$$\frac{P}{P^C} = 1 - \tfrac{3}{2}(\pi\varepsilon)^{2/3}$$

$$= 1 - 3.22\varepsilon^{2/3} \tag{5.29}$$

The arch with no initial pre-stress

Here $A_0 = -H$ as before, giving

$$P = P^C - \frac{3\ \pi^4 EIH}{2\ L^3}(6\pi\varepsilon)^{2/3}$$

Again non-dimensionalizing with respect to P^C, given here by equation (5.20), we have

$$\frac{P}{P^C} = 1 - \tfrac{3}{4}(6\pi^2\varepsilon^2)^{1/3}$$

$$= 1 - 2.92\varepsilon^{2/3} \tag{5.30}$$

5.9 COMPARISON WITH EXPERIMENTS

The careful series of experiments performed by Roorda in the 1960s included a buckled strut used as an arch,[53-6] and this provides convenient data for comparison with theoretical predictions. The bottom of Figure 5.5 shows experimental and theoretical imperfection-sensitivities with load off-set as imperfection, while the upper plot gives experimental equilibrium points for a 'near-perfect' system, together with the theoretical perfect response. We note that both plots are non-dimensionalized with respect to P^C, although experimentally this value is difficult to determine because of the infinite slope of the two-thirds power law cusp at C. The experimental points may thus be subject to some slight vertical shift, relative to the theoretical curves.

We see that the upper plot shows good agreement between experiment and theory over small displacements, but for the larger displacements included here

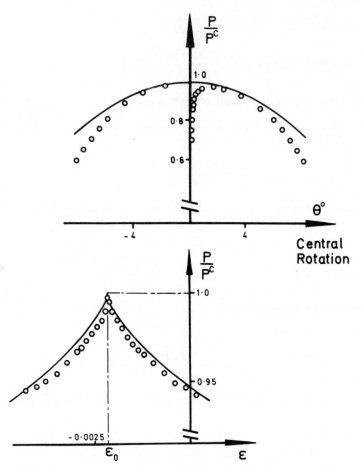

Figure 5.5 Comparison between the present theory and the experiments of Roorda[53]

the results are seen to diverge. This clearly indicates the range of accuracy of the asymptotic approach; agreement could doubtless be improved by extending the analysis to a higher order. Agreement is good in the lower plot for the full range of imperfections considered.

We see here that, with a controlled imperfection introduced by displacing the load a short distance ε from the centre-line of the arch, the cusp appears off-set an amount ε_0 as shown. This value is needed to counterbalance other inevitable manufacturing imperfections ξ, over which the experimentalist has no control, and bring the system back to a cusp bifurcation.

In theoretical terms[67] the $(\varepsilon + \xi)$ would be the overall major imperfection that is related to $(P - P^C)$ by a two-thirds power law. This then gives the off-set of $\varepsilon_0 = \xi$, when the experimentalist simply plots his control ε against P. A similar effect is observed in the next section, when tilt is introduced as a second controlled imperfection.

It is also of considerable interest to compare absolute values of P^C without the non-dimensionalization of Figure 5.5. Recently, a series of twelve tests on spring steel arches under different combinations of H and A_0 has been performed at Imperial College, London, by Heath, and the experimental P^C values thus obtained compared with the predictions of the present theory; for the latter, a value of bending stiffness EI was determined experimentally in a beam test, and was found to be significantly less than expected by taking E as $205\,kN/mm^2$, the normally accepted value for steel. A correction was also introduced into the theory, to allow for the point of load application being at a finite height h above the arch centre.

The results of the study were generally encouraging. There was a trend to more accuracy as both H and A_0 approached zero, which would be expected from our neglect of higher-order H/L type terms. For the lowest height considered, $H = 2.95\,cm$ with $A_0 = 0$ and $L = 61\,cm$, agreement is achieved to within 1%. With the larger values of H and A_0 considered, the experimental P^C consistently lay above the theoretical, suggesting that the higher-order terms, if included analytically, would carry a re-stabilizing effect. In all cases, agreement is achieved to within 10%. These results are further presented in graphical form in a recent publication.[82]

5.10 TILT AS A SECOND IMPERFECTION

We next briefly discuss a small tilt of the arch, relative to the direction of the load action, as a second imperfection. This has the effect of tilting the cusp geometry itself, as seen in the previous chapter; we are thus effectively introducing the third control parameter suggested by Golubitsky and Schaeffer for structural stability of the equilibrium paths.[6]

It is a relatively simple exercise in geometry to determine the horizontal displacement, Δ, of the crown of the arch ($x = L/2$), as the system moves through its buckling displacement Q_2,

$$\Delta = -\frac{2\pi H}{3 L}Q_2 + \frac{8}{3}\frac{\pi}{HL}Q_2^3 + \cdots \qquad (5.31)$$

Here, as before, higher terms in the coefficients are neglected using the moderate rise assumption $(H/L)^2 \ll 1$. Tilting the load action relative to the arch a small angle ϕ, positive in the clockwise sense of Figure 5.1, we thus have the potential function

$$V = U - P\cos\phi(q_1 - 2\pi\varepsilon Q_2) - P\Delta\sin\phi$$

adapted from equation (5.26). Expanding the trigonometric functions as power series in ϕ, we have

$$V = U - Pq_1 + 2\pi P\varepsilon Q_2 + \frac{2\pi H}{3 L}P\phi Q_2 - \frac{8}{3}\frac{\pi}{HL}P\phi Q_2^3 + \text{higher-order in } \phi \quad (5.32)$$

We now replace ε by a new imperfection γ, defined by

$$\varepsilon = \gamma - \frac{1}{3}\frac{H}{L}\phi$$

Substituting into the potential function, we obtain

$$V = U - Pq_1 + 2\pi P\gamma Q_2 - \frac{8}{3}\frac{\pi}{HL}P\phi Q_2^3 + \text{higher-order} \qquad (5.33)$$

This has the three control parameters γ, P, and ϕ acting on a linear, quadratic, and cubic form of Q_2, respectively, and is thus in the general form discussed in the previous chapter and by Golubitsky and Schaeffer.[6]

We note that $\gamma = 0$ defines a special direction in imperfection space, along which the initial two-thirds power law variation in critical load is lost. Here the first-order effects of the two imperfections ε and ϕ have effectively cancelled each other out, and we are determining the initial orientation of a line of cusps belonging to a cusped surface in $P-\varepsilon-\phi$ control space; slicing this surface at any orientation other than $\gamma = 0$, of course, gives the familiar two-thirds power law. As we emerge in the special direction $\gamma = 0$, the cusp defined by $\phi = \text{constant} \neq 0$ will itself tilt, as shown in Figure 4.15.

Experimental evidence for the existence of the cusped surface, and its tilt, has recently been obtained at Imperial College, London. Roorda has developed a similar approach in his treatment of first- and second-order imperfections,[55] where schematic representations of both the tilted cusp and its associated equilibrium paths can be found. However, his analysis is considerably complicated by the presence of coefficients such as $\partial^3 V/\partial P\partial\phi\partial Q_2|^C$ and $\partial^3 V/\partial\phi\partial Q_2^2|^C$, which Roorda retains in the interests of generality, but are absent from the present study. The first of these, which plays a key role in Roorda's theory, is clearly eliminated by the transformation to the ϕ-γ representation of imperfection space. The second can quite neatly be shown to vanish by symmetry considerations, as discussed in Chapter 8; we note that such symmetries are often present in systems displaying symmetric bifurcations.[81] (See also a recent paper by Thompson,[67] for a fundamental general treatment of major and minor imperfections in the cusp.)

5.11 CONVOLUTIONS IN THE SYMMETRIC RESPONSE

Finally, we briefly discuss the response when the arch is constrained to remain symmetrical about its vertical centre-line, and thus follows the non-linear fundamental path shown on the right of Figure 5.4. Symmetric arches have been studied theoretically by Biezeno and Grammel,[83] and numerically by Harrison;[84] excellent experimental results are presented by Croll and Walker.[85]

These studies all suggest a curious looping of the fundamental path on a plot of load against its corresponding deflection, illustrated here in Figure 5.6. The phenomenon is discussed in the light of stability under dead and rigid loading by

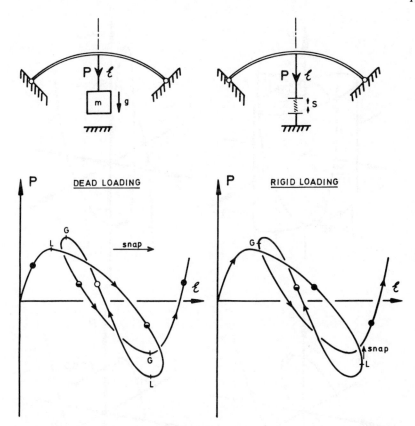

Figure 5.6 Load–corresponding deflection response of a symmetric arch (after Biexeno and Grammel, 1960). A solid circle denotes zero, a half-solid circle denotes one, and an open circle denotes two degrees of instability. Note that $\mathscr{E}$ here represents the corresponding deflection of the load (not the end-restraint, as in the rest of the chapter)

Thompson.[18] We explain it by reference to the equilibrium paths of Figure 5.7.

Here we assume that, under conditions of symmetry, the principal interaction arises between the first and third harmonics of the buckled strut, represented by the amplitudes Q_1 and Q_3. Moving from the top-left to the bottom-right of the figure we have a progressively decreasing extensional stiffness (EA) in comparison with the flexural stiffness EI.

At the top-left, the arch is completely inextensional. We thus have a constraint cylinder in $Q_1 - Q_3 - P$ space as shown, entirely analogous to the earlier cylinder of Figure 5.4. However, instead of bifurcating paths the response now comprises two entirely unconnected paths, each displaying a single limit point. Under dead-loading conditions, snap buckling is initiated at each limit point as shown, the system eventually stabilizing at the same load level on the other path. Under rigid loading, no snaps can take place.

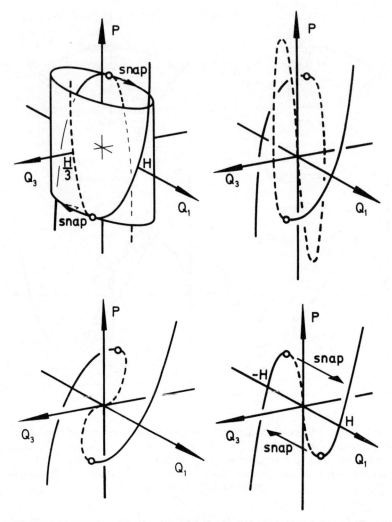

Figure 5.7 The effect on the symmetric response of reducing axial stiffness

In the diagram at the top-right, extensibility is allowed and the constraint cylinder disappears. However, the concept can still be useful, for we can imagine a set of concentric cylinders, decreasing in cross-sectional size with increasing axial shortening. Such reasoning allows the two paths to connect in their unstable regimes, generating two more limit points together with a doubly unstable equilibrium path passing through the origin, where it represents the fully compressed (straight) configuration of the arch. We note that the snap buckling shown in the top-left diagram can still arise in much the same way, but snaps are now also possible under conditions of rigid loading.

With increasing extensibility the second pair of limit points converge, finally resulting in a cut-off point critical with respect to Q_3, as shown at the bottom-left of Figure 5.7. This is merely an unfamiliar view of the cusp geometry associated with Euler buckling of a strut into its third harmonic, with end-constraint $\mathscr{E}$ and side-load P as control parameters. The whole phenomenon is, of course, unstable with respect to Q_1.

Further reducing the extensional stiffness eventually results in the situation at the bottom-right of Figure 5.7, in which Q_3 now has a minor part to play. The critical load required for buckling into the third harmonic cannot now be developed in the straight strut by applying the constraint $\mathscr{E}$. The response is qualitatively the same as the simple, one-degree-of-freedom, extensional arch of Section 4.1.

Assuming that the response shown at the top-right of Figure 5.7 is representative of a shallow, nearly inextensional arch, we obtain the qualitative form of Figure 5.6 on plotting load against its corresponding deflection. Further loops are added with increasing arch height as, in precisely analogous fashion to Q_3, modes Q_5, Q_7, etc. experience cut-off points in turn.

These ideas have recently been explored both experimentally and theoretically at Imperial College, London, by K. Jordan. We note finally that the response at the bottom-right of Figure 5.7 is, in fact, part of an underlying butterfly catastrophe, as demonstrated by Poston and Stewart using a link model.[5]

6

Interactive Buckling Phenomena

Interactive buckling problems with two coincident modes are controlled by the umbilic catastrophes. These, with two essential active coordinates q_1 and q_2, are more complex phenomena than those of the earlier chapters, and cannot reasonably be treated in the same depth. To allow us to describe them concisely, analysis is kept to a minimum, with reference to published sources or Chapter 7 where appropriate, and we rely on three-dimensional schematic diagrams and plots whenever possible. The behaviour is introduced via a deceptively simple two-degree-of-freedom guyed cantilever model due initially to Thompson and Gaspar,[86] and later extended by Hunt *et al.*[87] Further examples are provided by Hansen.[88]

One way of classifying two-fold bifurcations is by their symmetry properties; we have either *full asymmetry*, *semi-symmetry* (symmetry of one mode in the presence of the other), or *full symmetry* (symmetry of each mode in the presence of the other). However, the catastrophe theory classification is a little more subtle, the hyperbolic and elliptic umbilics embracing both of the first two categories above, the double cusp relating to the third, and the parabolic umbilic involving symmetry in each mode independently, but broken symmetry in the combination. We find that symmetry is an important physical property in many perfect structural systems, and is itself a generator of bifurcating behaviour, as discussed in Chapter 8. We thus give it strong emphasis here.

In many of the detailed analyses of the reference list, the potential function, instead of being written V, is represented by $\mathscr{A}$. This carries a special significance in the general formulation, specifically denoting a diagonalized function, expanded as a Taylor series about a single-valued fundamental path, from which $n - m$ passive coordinates have been eliminated; we use it in the more analytical context of Chapter 7, for instance. Here, although we always suppose we have such a specialized function, we remain with the notation V, keeping consistent with the earlier chapters and many associated studies.

6.1 THE GUYED CANTILEVER

The system under consideration comprises a light rigid rod of length L, pinned at its base, and supported by three linear springs as shown in Figure 6.1. It carries a

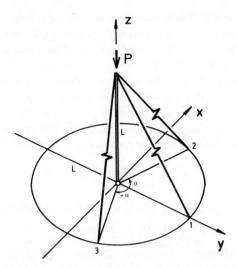

Figure 6.1 The simple guyed cantilever model

dead vertical load P, and with the rod vertical all the springs have an inclination of 45 degrees to the horizontal. One spring is initially in the plane yz, and the other two are placed symmetrically about it, each at an angle of α measured in the base plane as shown.

Buckling can take place in two distinct modes, one in the plane yz and the other initially at right-angles to it, and by adjusting the spring stiffnesses these modes can be obliged to buckle simultaneously at the same critical load; the complete symmetry of the system about the plane yz ensures that this gives a semi-symmetric point of bifurcation. By adjusting α the system can be scanned through a complete family of such points, and this was related in the original study[86] to a symmetric sweep through the triangular hypocycloid section of Zeeman's celebrated *umbilic bracelet*, described in full in his book;[4] we return to this point later.

Potential function and eigenvalue analysis

Displacements of the system can be fully described by the x and y coordinates of the top of the rod, and we thus introduce the non-dimensional generalized coordinates

$$Q_1 = x/L, \quad Q_2 = y/L$$

Imperfect systems can be defined by allowing these coordinates to take the initial values $Q_1 = Q_1^0$ and $Q_2 = Q_2^0$ under zero load and with the springs unstressed. For the perfect system $Q_1^0 = Q_2^0 = 0$.

In the initial study[87] it was assumed that a given amount of spring stiffness, K, was available, and this was distributed among the springs in some way to ensure complete coincidence of the two contributing bifurcations. Thus if we write the stiffness of the equal second and third springs as $c_2 = \gamma K$, then the stiffness of the first spring must be $c_1 = (1 - 2\gamma)K$, and there must be a unique value of γ for each initial configuration of the system α. Later,[88] although the system was restricted to a total available stiffness as before, γ was allowed to vary while α remained fixed, so the two contributing bifurcations could be separated in a controlled fashion. Here we follow the second, more general, formulation.

We now write down the potential function, which is derived in full in reference 87. It will be sufficient for our purposes to take the form

$$
\begin{aligned}
S = {} & (\tfrac{1}{8}c_1 L^2 + \tfrac{1}{4}c_2 L^2 \cos^3 \alpha)Q_2^3 + \tfrac{3}{4}c_2 L^2 \sin^2 \alpha \cos \alpha Q_1^2 Q_2 \\
& + (\tfrac{1}{4}c_1 L^2 + \tfrac{1}{2}c_2 L^2 \cos^2 \alpha)Q_2^2 + \tfrac{1}{2}c_2 L^2 \sin^2 \alpha Q_1^2 \\
& - \tfrac{1}{2}PL(Q_1^2 + Q_2^2) - (\tfrac{1}{2}c_1 L^2 + c_2 L^2 \cos^2 \alpha)Q_2^0 Q_2 \\
& - c_2 L^2 \sin^2 \alpha Q_1^0 Q_1 + \text{higher-order terms}
\end{aligned}
\tag{6.1}
$$

We note that the absence of the quadratic cross-term $Q_1 Q_2$, guaranteed by the semi-symmetry, gives a diagonalized potential function; in the earlier studies the Q_i are written u_i in recognition of this. The semi-symmetry also ensures the absence of the cubic Q_1^3 and $Q_1 Q_2^2$ terms.

For the perfect system with $Q_1^0 = Q_2^0 = 0$ we have a fundamental equilibrium solution given by $Q_1 = Q_2 = 0$ for all c_1, c_2, and P. Critical equilibrium states on this trivial fundamental path F can be found by differentiating twice, evaluating on the path, and setting the resulting stability coefficients S_{11}^F and S_{22}^F to zero. This gives the two critical loads

$$
P_{\text{crit}}^1 = c_2 L \sin^2 \alpha = \gamma K L \sin^2 \alpha
$$
$$
P_{\text{crit}}^2 = \tfrac{1}{2}c_1 L + c_2 L \cos^2 \alpha = \tfrac{1}{2}KL - \gamma KL \sin^2 \alpha
$$

associated with the modes Q_1 and Q_2, respectively.

Let us first consider complete coincidence of the critical loads. Equating P_{crit}^1 and P_{crit}^2 we obtain

$$
\gamma^C = \frac{1}{4 \sin^2 \alpha}
$$

$$
P^C = \tfrac{1}{4}KL
$$

where C is the two-fold compound branching point.

We next introduce an incremental *loading parameter*, p, and an incremental *splitting parameter*, Γ, defined by

$$
p = P - P^C
$$
$$
\Gamma = \gamma - \gamma^C
$$

We see that varying Γ separates the two contributing bifurcations on the fundamental path.

Substituting into the potential function and dividing through by

$P^C L = K L^2/4$, a positive constant, we obtain

$$V = \frac{4S}{KL^2} = \frac{1}{4}\left(2 - \frac{1}{\sin^2\alpha} + \frac{\cos^3\alpha}{\sin^2\alpha}\right)Q_2^3 + \tfrac{3}{4}\cos\alpha Q_1^2 Q_2 - \tfrac{1}{2}\frac{p}{P^C}(Q_1^2 + Q_2^2)$$

$$+ 2\Gamma(Q_1^2 - Q_2^2)\sin^2\alpha - Q_2^0 Q_2 - Q_1^0 Q_1 + \text{higher-order terms} \qquad (6.2)$$

We now have a non-dimensionalized form of the potential function, expanded about the compound critical state C, which can be directly related to the general form discussed below. We note that the equilibrium equations are unchanged by non-dimensionalizing in this way, but if we had divided through by a *negative* constant, a *maximum* of the resulting function would then imply stability.

Parametric sweep through the umbilic bracelet

The catastrophe theory classification of the umbilics rests solely on the cubic form of the potential function at complete coincidence of the two contributing bifurcations, which is determined by setting all controls, p, Γ, Q_1^0, and Q_2^0, to zero. For semi-symmetry this becomes

$$V = \tfrac{1}{6}V_{222}^C Q_2^3 + \tfrac{1}{2}V_{112}^C Q_1^2 Q_2 \qquad (6.3)$$

and we note that it contains just two coefficients instead of the four of the general

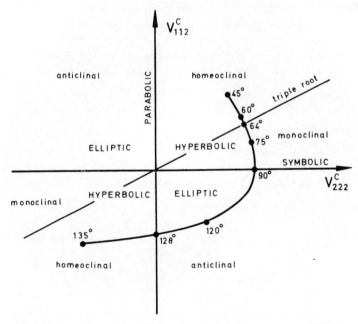

Figure 6.2 The path taken by varying α through the space of the semi-symmetric cubic coefficients

cubic form. By varying α we can scan through the space of all semi-symmetric forms, as shown in Figure 6.2.

The figure indicates the catastrophe theory classification for semi-symmetric branching into hyperbolic or elliptic umbilics, depending on whether the coefficients V_{222}^C and V_{112}^C are the same or opposite in sign. The axes, $V_{222}^C = 0$ and $V_{112}^C = 0$, represent the higher-order parabolic and symbolic umbilics, respectively. The diagram also indicates the bifurcational subclassification of the umbilics, in which the elliptic umbilic relates directly to anticlinal branching, but the hyperbolic umbilic further subdivides into two quite different forms called monoclinal and homeoclinal branching.

For the general cubic form, the space of the coefficients corresponding to Figure 6.2 is clearly four-dimensional. However, by means of a simple transformation of variables and subsequent projection,[4] Zeeman views the behaviour in three dimensions, via the concept of the *umbilic bracelet*; a model of its geometry, beautifully carved in hardrock maple by Tim Poston, is shown in Plate 6.1. We see that the bracelet is a doughnut shape, exhibiting on its surface a single line of cusps, which passes three times around the ring of the doughnut while traversing

Plate 6.1 The umbilic bracelet of Zeeman,[4] hand-carved by T. Poston, and reproduced with permission

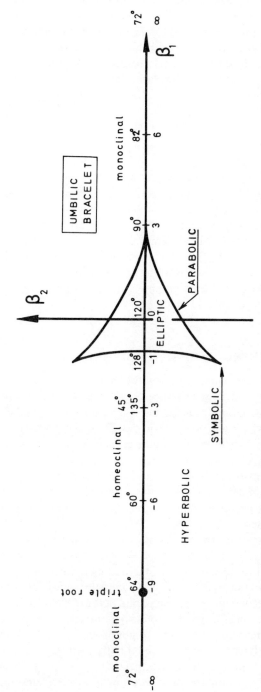

Figure 6.3 The path taken through Zeeman's umbilic bracelet by varying α. Semi-symmetry coincides with the β_1 axis

its circumference once; the ring thus has a triangular hypocycloid section. Cubic forms relating to the elliptic umbilic catastrophe appear inside the bracelet, hyperbolic umbilic points appear outside, parabolic umbilic points are on the smooth boundary, and symbolic umbilics on the cusps.

A section through the umbilic bracelet is shown in Figure 6.3. This plots two of the three new variables after transformation and projection, the third being held constant. For semi-symmetry, we have the special results,[86]

$$\beta_1 = \frac{3V^C_{222} + 3V^C_{112}}{V^C_{222} - 3V^C_{112}}, \qquad \beta_2 = 0$$

and the system is restricted to the β_1 axis.

Increasing α from 45 degrees, we find that β_1 decreases from -3, until at $\alpha = 72$ degrees, $V^C_{222} = 3V^C_{112}$ and $\beta_1 = \pm \infty$. Increasing α further, β_1 decreases from $+\infty$, and the system meets the bracelet at one of its cusps with $\alpha = 90$ degrees, and leaves it at $\alpha = 128$ degrees. Finally, at $\alpha = 135$ degrees we return to the point represented by $\alpha = 45$ degrees, because these two limiting values yield identical structures with a vanishing first spring ($\gamma = \frac{1}{2}$ so $c_1 = 0$). We see that the guyed cantilever model thus gives a scan through the entire β_1 axis of the umbilic bracelet.

6.2 SEMI-SYMMETRIC POINTS OF BIFURCATION

We now subject the semi-symmetric points of bifurcation to general examination, presenting special results for the simple guyed cantilever where appropriate. Analytical details are largely omitted, but can be found at the end of Section 7.2 of the next chapter, where semi-symmetric branching is used as an example of the general bifurcation algorithm. This culminates in Table 7.1, listing ten separate analyses, each of which relates to some different aspect of compound bifurcation; examples include post-buckling, imperfection-sensitivity, and loci of secondary bifurcations, at complete and near-coincidence of the two contributing bifurcations. Low-order perturbation equations are given, and a method of solution indicated in all cases.

Here we concentrate on the results of such a detailed survey. The phenomena are introduced largely in order of complexity, as the control parameters are included in turn. Thus we start with just a loading parameter Λ and examine the equilibrium paths of the perfect system at complete coincidence of the contributing bifurcations. Second, imperfection parameters are included and the imperfect equilibrium paths discussed. Next, a splitting parameter σ is used to separate the contributing bifurcations and the equilibrium paths re-examined. The imperfection-sensitivity is then discussed, finally with the four controls of load, two imperfections, and a splitting parameter.

We note that semi-symmetric bifurcation further subdivides into monoclinal, anticlinal, and homeoclinal branching, each of which is treated here in full.

Equilibrium paths of the perfect system

We suppose that, after the elimination of passive coordinates, the perfect system is described by the two-degree-of-freedom potential function

$$V(u_i, \Lambda) = \tfrac{1}{6} V^C_{222} u_2^3 + \tfrac{1}{2} V^C_{112} u_1^2 u_2$$
$$+ \tfrac{1}{2}(\Lambda - \Lambda^C)(V'^C_{11} u_1^2 + V'^C_{22} u_2^2) + \text{higher-order terms} \qquad (6.4)$$

Here the u_i represent *principal* generalized coordinates, the potential function having been diagonalized by the semi-symmetry, Λ is a loading parameter, and we have complete coincidence of the two contributing bifurcations at the compound critical state C.

A post-buckling study using this potential function is presented in our earlier book,[1] giving the three solutions for the post-buckling equilibrium paths in rate space

$$\left.\frac{u_1^{(1)}}{u_2^{(1)}}\right|^C = 0, \qquad \left.\frac{\Lambda^{(1)}}{u_2^{(1)}}\right|^C = -\left.\frac{V_{222}}{2V'_{22}}\right|^C,$$

$$\left.\frac{u_1^{(1)}}{u_2^{(1)}}\right|^C = \pm \sqrt{\left.\left(\frac{2V'_{22}}{V'_{11}} - \frac{V_{222}}{V_{112}}\right)\right|^C}, \qquad \left.\frac{\Lambda^{(1)}}{u_2^{(1)}}\right|^C = -\left.\frac{V_{112}}{V'_{11}}\right|^C \qquad (6.5)$$

where the bracketed superscript refers to full differentiation with respect to some unspecified perturbation parameter s. These solutions become the path tangents when mapped directly into the space u_i–Λ.

The first solution is identical to the slope of the single post-buckling path for the distinct asymmetric point of bifurcation critical with respect to u_2 (equation (4.11) of Chapter 4). It is thus clear that this path is unaffected by the separation of the two contributing bifurcations on the fundamental path,[89] as seen later; we term it the *uncoupled* post-buckling path. The other two *coupled* paths may or may not exist, giving rise to a temporary classification into single- or three-path semi-symmetric branching; this is further subdivided later by the nature of the imperfection-sensitivity. We note that in the three-path case where the condition

$$\left.\frac{2V'_{22}}{V'_{11}}\right|^C > \left.\frac{V_{222}}{V_{112}}\right|^C \qquad (6.6)$$

is met, the two coupled paths lie symmetrically opposed about the u_2 axis in u_i–Λ space.

Clearly, it would be a simple matter to substitute our known coefficients for the guyed cantilever, namely,

$$V^C_{222} = \tfrac{3}{2}\left(2 - \frac{1}{\sin^2 \alpha} + \frac{\cos^3 \alpha}{\sin^2 \alpha}\right), \qquad V^C_{112} = \tfrac{3}{2}\cos\alpha, \qquad V'^C_{11} = V'^C_{22} = -1$$

and obtain the post-buckling path tangents for the problem. However, this adds little to our general description and so is omitted. Also, we leave schematic drawings of the paths until later; these are included in the figures of the next two subsections.

Equilibrium paths of imperfect systems

Equilibrium paths for the complete family of imperfect systems, generated by varying the two imperfections, are shown for the three semi-symmetric points of bifurcation in Figures 6.4–6.6. In each case the paths of the perfect system form a

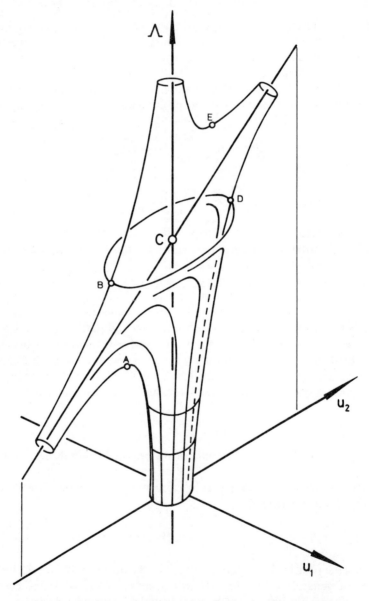

Figure 6.4 Perfect and imperfect equilibrium paths for the monoclinal point of bifurcation

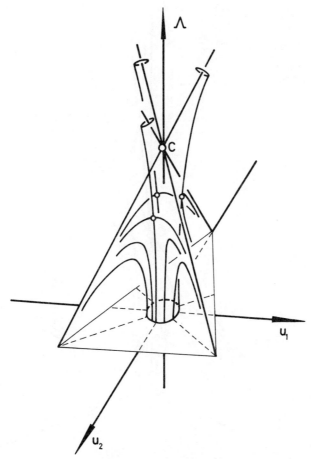

Figure 6.5 Perfect and imperfect equilibrium paths for the
anticlinal point of bifurcation

skeleton for a solid tree of imperfect paths, with the trunk fully enclosing the
fundamental path and the branches fully enclosing the post-buckling path or paths.
The monoclinal point of bifurcation of Figure 6.4 is merely the single path case of
the previous subsection. In Figures 6.5 and 6.6 we see the difference between
anticlinal and homeoclinal branching; for the former, the coupled and uncoupled
paths fall in opposite directions with respect to the u_2 axis, while for the
homeoclinal case the reverse is true.

   The imperfect equilibrium paths have themselves been subjected to general
study,[90,91] but we use no such results here, so the diagrams are schematic
representations and not plotted. Analysis of such paths is rendered awkward by
the non-linearities, and is unnecessary as far as the imperfection-sensitivity
analysis is concerned; we can pinpoint critical equilibrium states on a path
without having to solve for the complete path directly.

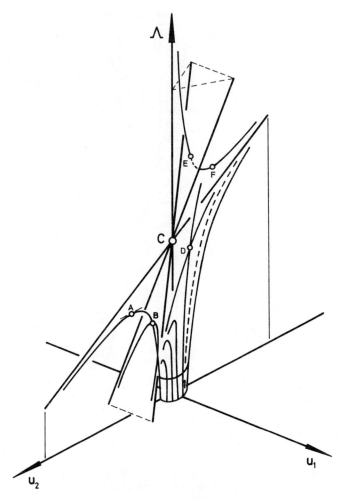

Figure 6.6 Perfect and imperfect equilibrium paths for the homeoclinal point of bifurcation

Equilibrium paths at near-coincidence

If we replace the imperfection parameters with a splitting parameter, which, like
the load, acts on a quadratic instead of a linear form of the coordinates in the
potential function expansion, then the patterns of the equilibrium paths are
broken up in a somewhat different fashion. Let us thus take the potential function

$$V = \tfrac{1}{6}V^C_{222}u_2^3 + \tfrac{1}{2}V^C_{112}u_1^2 u_2 + \tfrac{1}{2}(\Lambda - \Lambda^C)(V'^C_{11}u_1^2 + V'^C_{22}u_2^2)$$
$$+ \tfrac{1}{2}\sigma V'^C_{11}u_1^2 + \text{higher-order terms}$$

We see that the splitting parameter σ, when introduced in this way, gives a direct
measure of separation on the fundamental path of a fixed asymmetric bifurcation

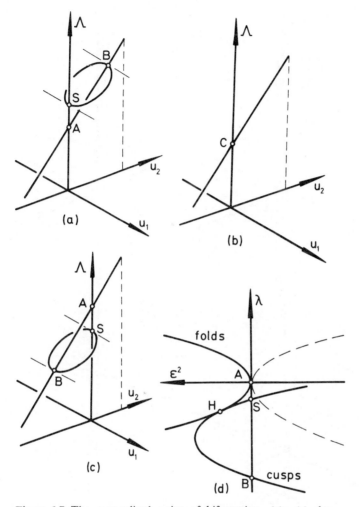

Figure 6.7 The monoclinal point of bifurcation: (a) – (c) show patterns of equilibria for a perfect system with $\sigma < 0$, $\sigma = 0$, and $\sigma > 0$, respectively; (d) shows the failure locus (imperfection-sensitivity) with the addition of a symmetry-preserving imperfection ε^2

A, critical with respect to u_2, and a shifting symmetric bifurcation S, critical with respect to u_1; these are the contributing bifurcations of the interaction.

The forms of the equilibrium paths with changing σ for the three semi-symmetric points are illustrated by the first three diagrams in each of Figures 6.7–6.9; again these are schematic representations of the behaviour. In each case we can see the appearance of A and S on the fundamental path, and an associated secondary bifurcation B on the uncoupled path.[92] These patterns were sketched by Chilver in 1967.[89]

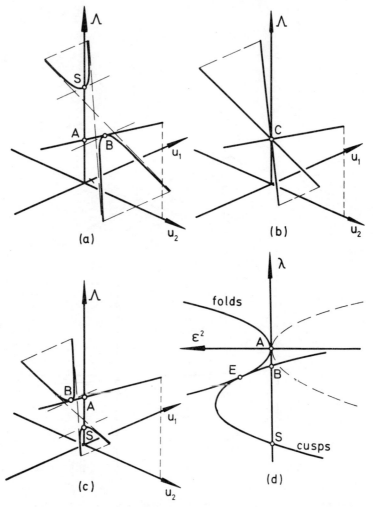

Figure 6.8 The anticlinal point of bifurcation. Patterns of equilibria for a perfect system and imperfection-sensitivity with a symmetry-preserving imperfection ε^2

Secondary bifurcations

None of the patterns of equilibria of Figures 6.4–6.9 can exist without some paths, other than the fundamental, exhibiting distinct bifurcations. These we term *secondary bifurcations*, and note that they may be either symmetric or asymmetric in form. Examples can be found in all the figures, specifically labelled *B* in some cases.

In an imperfection-sensitivity study, for example, secondary bifurcations can be pinpointed from among all possible critical states by the addition of an extra equation; we see this in Chapter 7. At complete coincidence, each of the post-

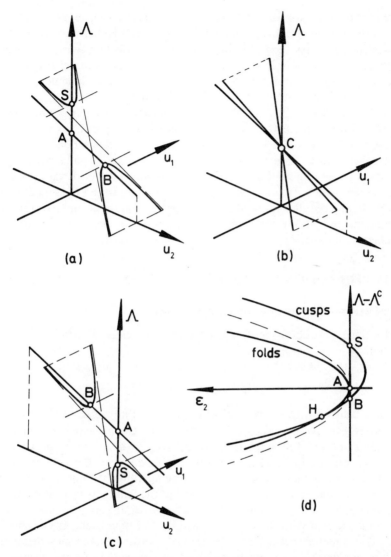

Figure 6.9 The homeoclinal point of bifurcation. Patterns of equilibria for a perfect system and imperfection-sensitivity with a symmetry-preserving imperfection ε^2

buckling paths of the perfect system is directly related to a locus of secondary bifurcations in the u_i–Λ space of Figures 6.4–6.6, as recognized by Johns;[90] however, the exact nature of this link remains rather subtle.[93]

The symmetric section

Before examining the full imperfection-sensitivity it is instructive to observe the behaviour of the system under a single *principal* imperfection ε^2, purely in the mode u_2. This has a special *symmetry-preserving* effect, since, unlike ε^1, it does not destroy the symmetry of the potential function in u_1. Whether we are considering the system geometry, equilibrium paths, or imperfection-sensitivity, we can always isolate behaviour in a special plane of symmetry, or *symmetric section*, of the complete behaviour. Examples of symmetric sections include the plane yz of Figure 6.1, and the u_2–Λ planes of Figures 6.4–6.9.

Analysis of the fully-coincident[93] and nearly-coincident[92] imperfection-sensitivity on the symmetric section, as outlined in Chapter 7, gives the plots at the lower right of Figures 6.7–6.9. In each case we have a parabolic line of cusps, corresponding to a symmetric bifurcation into the mode u_1, and a parabolic line of folds, representing limiting behaviour in the mode u_2. With variation in the splitting parameter σ the line of cusps rolls smoothly round the line of folds as shown, shifting the elliptic umbilic point E, or the hyperbolic umbilic point H, off the fundamental path to some imperfect system. These diagrams were originally sketched by Supple.[94]

6.3 IMPERFECTION-SENSITIVITY SURFACES

Imperfection-sensitivity at complete coincidence

For the full imperfection-sensitivity at complete coincidence we take the potential function

$$V = \tfrac{1}{6}V^C_{222}u_2^3 + \tfrac{1}{2}V^C_{112}u_1^2 u_2 + \tfrac{1}{2}(\Lambda - \Lambda^C)(V^{'C}_{11}u_1^2 + V^{'C}_{22}u_2^2)$$

$$+ V^{1C}_1 \varepsilon^1 u_1 + V^{2C}_2 \varepsilon^2 u_2 + \text{higher-order terms}$$

We note that both the cross-derivatives $V^{jC}_i (i \neq j)$ are zero, so the ε^i are thus defined as *principal imperfections* associated with the principal coordinates u_i.

Analysis of the full imperfection-sensitivity,[93] as outlined in Chapter 7, involves defining all critical states about C in parametric form, and determining their loci on a general imperfection ray emerging from C. Of course, this includes all complementary path behaviour, as well as the failure locus of usual interest, encountered first during a natural loading sequence; the latter is usually, but not always, associated with the lowest critical load of the system. It was shown by Ho[95] that under a special normalization of coordinates and imperfections the most severe imperfection-sensitivity is directly associated with the steepest post-buckling path of the perfect system; an alternative proof is offered by Samuels.[96,97]

The results of a typical imperfection-sensitivity analysis are plotted in Figure 6.10 for a monoclinal point of bifurcation. This shows the full failure locus in the three-dimensional, load-imperfection, control space. The outer rim of each surface marks its intersection with a vertical hollow cylinder, and the radial lines on the surfaces correspond to variations along specific imperfection rays.

The topological form of Figure 6.10 can be recognized as the universal unfolding in control space of the hyperbolic umbilic catastrophe, so the three controls of load and two imperfections are sufficient for structural stability

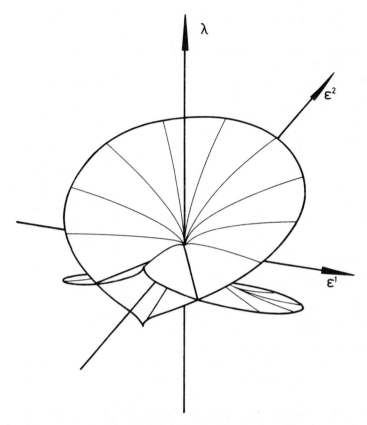

Figure 6.10 Full imperfection-sensitivity for the monoclinal point of bifurcation at complete coincidence, giving the universal unfolding of the hyperbolic umbilic catastrophe: here $\lambda = \Lambda - \Lambda^C$

according to Thom's criterion. However, in the refined sense of Golubitsky and Schaeffer, which recognizes the special role of the loading parameter,[6] extra controls are required; this is investigated no further here. The structural stability of the form allows transformations between different sets of controls and different systems to lend the imperfection-sensitivity plots greater generality;[87] this is seen in the next few subsections, where the different forms for the three semi-symmetric points are taken in turn.

Imperfection-sensitivity at the homeoclinal point of bifurcation

We start the detailed study of imperfection-sensitivity with homeoclinal branching, shown earlier in Figures 6.6 and 6.9. With the full four control parameters we are dealing with the potential function

$$V = \tfrac{1}{6}V^C_{222}u_2^3 + \tfrac{1}{2}V^C_{112}u_1^2u_2 + \tfrac{1}{2}(\Lambda - \Lambda^C)(V'^C_{11}u_1^2 + V'^C_{22}u_2^2)$$
$$+ \tfrac{1}{2}\sigma V'^C_{11}u_1^2 + V^{1C}_1\varepsilon^1u_1 + V^{2C}_2\varepsilon^2u_2 + \text{higher-order terms} \qquad (6.7)$$

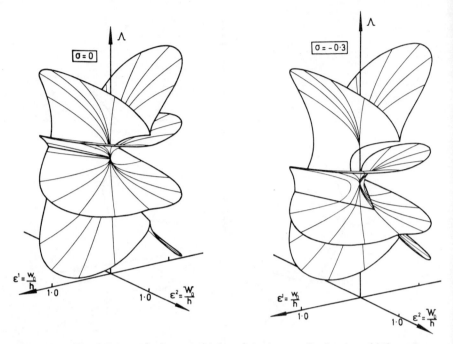

Figure 6.11 The full imperfection-sensitivity of the homeoclinic point of bifurcation at complete coincidence and with separation of critical points. Both exhibit the unique topology of the hyperbolic umbilic catastrophe

and we remember first, that condition (6.6) is upheld and second, that V^C_{222} and V^C_{112} must be of the same sign.

Using the full analysis,[93] described in Chapter 7, we obtain the imperfection-sensitivity surfaces of Figures 6.11 and 6. 12. The left-hand side of each gives a view of the failure locus at complete coincidence, while the right-hand sides give two perturbed versions of the locus at different but constant values of σ, as indicated. Each of these plots exhibits the unique topology of the hyperbolic umbilic catastrophe, structural stability ensuring that the form is preserved with variations in σ.

These plots have been obtained from a specific potential function for an interactive buckling problem in a stiffened plate, due to Tvergaard.[98] Here a wide plate is compressed in the direction of the stiffeners, the longitudinal edges remaining free. Two critical modes of buckling arise at or near the same load, an overall Euler mode, and a local plate buckling mode in which the stiffeners remain undeformed; for a detailed discussion on the phenomenon of interactive buckling in stiffened structures we refer to Chapter 8. We shall see later one way in which these plots can be reinterpreted for the simple guyed cantilever, and note that an alternative view involving 'critical imperfection territory' has been provided recently by Gaspar.[99]

A few general remarks on the nature of the right-hand plots are in order. First,

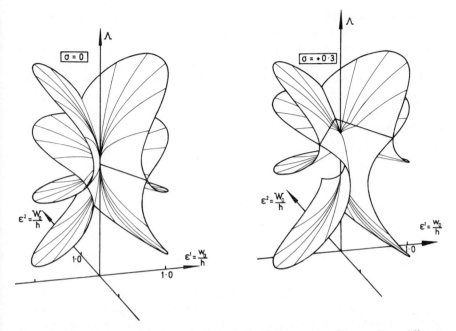

Figure 6.12 Back views of the hyperbolic umbilic of the previous figure, with a different value of the splitting parameter σ for the right-hand side

it should be noted that a polar representation of the imperfection parameters is employed in the analysis,[93] as suggested earlier. Second, a polar representation of the Λ–σ space is also central to the analysis when $\sigma \neq 0$, and each plot is developed in its entirety from the fundamental imperfection-sensitivity relation,[92]

$$\lambda^* = A\varepsilon^{1/2}$$

where ε is the radial imperfection and λ^* is the radial 'generalized load' in Λ–σ space. Now ε, λ^*, and A all vary over the surface, but we could multiply λ^* by some factor m while multiplying ε by m^2, and obtain exactly the same equation as before for any point on the surface. In this way we can rescale both the full Λ–σ 'load' space and the full imperfection space ε^i, and thus, for example, reinterpret the plots of Figure 6.11 and 6. 12 in a much more localized sense. This rescaling of the Λ-axis must, however, affect the position of the critical point with respect to the origin, so we must replace Λ by $\Lambda - \Lambda_k$ to allow for a shift of origin.

Thus we can, for example, interpret Figure 6.11 as it stands for the imperfection-sensitivity of the Tvergaard stiffened plate, with $\Lambda_k = 0$ and the two values of $\sigma = 0$ and $\sigma = -0.3\Lambda^C$, for the left and right plots, respectively. Alternatively, we could set $m = 10$, say, and thus replace Λ by 10Λ, σ by 10σ, and each of the ε^i by $100\varepsilon^i$ to obtain a more localized view of the imperfection-sensitivity; the two plots are then valid for $\sigma = 0$ and $\sigma = -0.03\Lambda^C$, respectively, and we must have $\Lambda_k = 9\Lambda^C$ to correct for the position of the origin. In this way we can use the plots to give the imperfection-sensitivity of the Tvergaard plate for any reasonable value of σ, positive or negative.

But the plots can also be used for other structural systems, for example, the guyed cantilever of Section 6.1. Here simple linear transformations can be derived[88] which link the control parameters of the cantilever model with those of the general formulation. Thus, at the special starting condition of $\alpha = 130.69$ degrees, which Figures 6.2 and 6.3 confirm as belonging to a region of homeoclinal branching, the left-hand plots of Figures 6.11 and 6.12 can be reinterpreted by setting

$$\frac{P}{P^C} = 2.90 \frac{\Lambda}{\Lambda^C}, \quad \Gamma = 0, \quad \varepsilon^1 = -0.297Q_1^0, \quad \varepsilon^2 = 1.02Q_2^0 \qquad (6.8)$$

with $\Lambda_k = -0.655\Lambda^C$. Alternatively, we could obtain a more localized view by setting

$$\frac{P}{P^C} = 0.290 \frac{\Lambda}{\Lambda^C}, \quad \Gamma = 0, \quad \varepsilon^1 = -29.7Q_1^0, \quad \varepsilon^2 = 102Q_2^0 \qquad (6.9)$$

with $\Lambda_k = 2.45\Lambda^C$, by the simple rescaling of axes discussed above. Clearly, the localized view gives a more accurate prediction of the imperfection-sensitivity since contamination from higher-order terms is necessarily much reduced.

For the right-hand diagrams with $\sigma \neq 0$ the relationships between P and Λ, ε^1 and Q_1^0, and ε^2 and Q_2^0, can be left unchanged, but clearly now $\Gamma \neq 0$, and the value of Λ_k must be altered. Thus for the right-hand side of Figure 6.11 we can use the results (6.8) but with

$$\Gamma = 0.189, \quad \Lambda_k = -0.805\Lambda^C$$

or for an example of a more localized view we can take the results (6.9) but with

$$\Gamma = 0.0189, \quad \Lambda_k = 2.30\Lambda^C$$

Similarly, for the right-hand side of Figure 6.12 we can use equations (6.8) but with

$$\Gamma = -0.189, \quad \Lambda_k = -0.505\Lambda^C$$

or for the localized view we take equations (6.9) with

$$\Gamma = -0.0189, \quad \Lambda_k = 2.60\Lambda^C$$

Imperfection-sensitivity at the anticlinal point of bifurcation

Having considered the first type of semi-symmetric branching in some depth we can deal more swiftly with the entirely analogous situations which exist for the other two. For anticlinal branching of Figures 6.5 and 6.8 we have the same potential function (6.7) as before, but with V_{222}^C and V_{112}^C of opposite sign; here condition (6.6) is automatically satisfied, since for an initially stable fundamental path, $V_{11}^{\prime C}$ and $V_{22}^{\prime C}$ must both be negative.

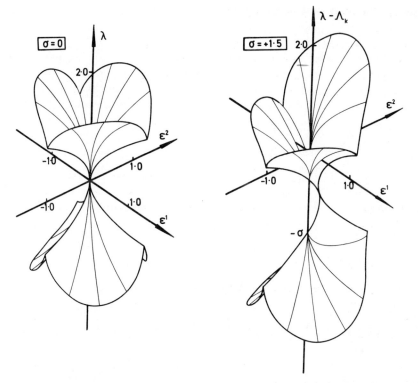

Figure 6.13 The full imperfection-sensitivity of the anticlinal point of bifurcation at complete coincidence and with separation of critical points. Both exhibit the unique topology of the elliptic umbilic catastrophe

Taking the particular potential function

$$V = u_2^3 - 3u_1^2 u_2 - \lambda(u_1^2 + u_2^2) - \sigma u_1^2 + \varepsilon^1 u_1 + \varepsilon^2 u_2 + \text{higher-order terms}$$

we obtain the plots of Figure 6.13 for the full imperfection-sensitivity. These both exhibit the unique topology of the elliptic umbilic catastrophe, and structural stability again ensures that the form is preserved with variation of σ. We see that this form differs in nearly every respect from the earlier plots of the hyperbolic umbilic, despite the similarities in their patterns of equilibria.

The left-hand version of the failure locus exhibits a clear tri-symmetry, and we note that most examples of the elliptic umbilic in physical systems arise because of some inherent tri-symmetry.[31,86] The presence of σ, of course, destroys this symmetry. We can once again reinterpret the right-hand plot for any appropriate value of σ, positive or negative, with a simple rescaling of the axes; the constant Λ_k is here explicitly included to allow for a later shift of origin.

Again, we can interpret the plots of Figure 6.13 specifically for the guyed cantilever of Section 6.1.[87] For the starting condition of $\alpha = 120$ degrees, which figures 6.2 and 6.3 confirm as being in a region of anticlinal branching and which

132

exhibits the tri-symmetry discussed above, we set

$$p = \tfrac{1}{4}\lambda P^C, \quad \Gamma = 0, \quad \varepsilon^1 = -8Q_1^0, \quad \varepsilon^2 = -8Q_2^0$$

for the left-hand side. We find that the use of an incremental loading parameter, λ, has meant that here we are not obliged to allow for a shift of origin, as previously. The appropriate zero-load position on the plot can be found by setting $p = -P^C$, giving $\lambda = -4$. For the right-hand plot we have the same results, except now

$$\Gamma = -0.0625, \quad \Lambda_k = 0.75$$

and we see that now we must have an origin shift. Clearly, we can rescale the axes to give a more (or less) localized view, as before.

Imperfection-sensitivity at the monoclinal point of bifurcation

For monoclinal branching of Figures 6.4 and 6.7 we again have the same potential function (6.7), this time with V_{222}^C and V_{112}^C of the same sign and condition (6.6) violated. Taking the particular potential function

$$V = \tfrac{17}{6}u_2^3 + 3u_1^2 u_2 - \lambda(u_1^2 + u_2^2) - \sigma u_1^2 + \varepsilon^1 u_1 + \varepsilon^2 u_2 + \text{higher-order terms}$$

we obtain the plots of Figure 6.14 for the full imperfection-sensitivity. Here we have once again the unique topology of the hyperbolic umbilic catastrophe for

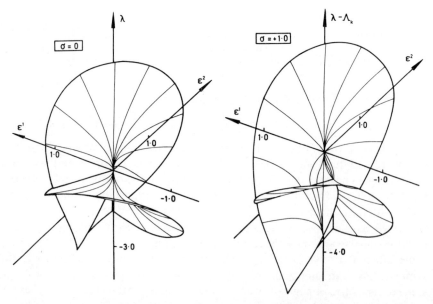

Figure 6.14 The full imperfection-sensitivity of the monoclinal point of bifurcation at complete coincidence and with separation of critical points. Here we have the back view of the form of Figure 6.10, a different manifestation of the hyperbolic umbilic from Figures 6.11 and 6.12

both zero and non-zero values of σ, and the right-hand plot can be reinterpreted for any σ value by a simple rescaling of axes.

Again, we can interpret the plots of Figure 6.14 specifically for the guyed cantilever model.[87] For the starting condition of $\alpha = 71.04$ degrees, which Figures 6.2 and 6.3 confirm as being in a region of monoclinal branching, we set

$$p = 0.162\lambda P^C, \quad \Gamma = 0, \quad \varepsilon^1 = -12.31 Q_1^0, \quad \varepsilon^2 = -12.31 Q_2^0$$

for the left-hand side. Again we need no shift of origin. Setting $p = -P^C$ to find the zero-load position, we have $\lambda = -6.16$ on the plot. For the right-hand side we have the same results, except now

$$\Gamma = -0.0227, \quad \Lambda_k = 0.5$$

and again we must have a shift of origin. As before, we can rescale the axes for a more localized view.

6.4 FULLY ASYMMETRIC POINTS OF BIFURCATION

The subclassification of semi-symmetric branching into monoclinal, anticlinal and homeoclinal forms can be neatly extended to the fully asymmetric case by a Lagrange-multiplier argument.[13] Here, as is to be expected from the bifurcational formalism, the loading parameter Λ has a special role to play. We note that, compared with the semi-symmetric form, fully asymmetric two-fold branching is a rare phenomenon in structural mechanics.

It is the central theme of both Koiter's pioneering contribution[12] and catastrophe theory that the topology of the potential function at the most-singular state, $V(q_i, \Lambda^C, 0)$, determines the initial post-buckling and associated phenomena. Thus, about a fully asymmetric two-fold branching point C, the behaviour is initially dependent upon the cubic form

$$V = \tfrac{1}{6} V_{111}^C q_1^3 + \tfrac{1}{2} V_{112}^C q_1^2 q_2 + \tfrac{1}{2} V_{122}^C q_1 q_2^2 + \tfrac{1}{6} V_{222}^C q_2^3 \tag{6.10}$$

We note that the catastrophe theory classification of the umbilics depends on the root structure of this cubic form; setting the form to zero gives either one or three real distinct solutions q_1/q_2, the two cases which correspond to the hyperbolic and elliptic umbilics, respectively. The solutions, of course, represent routes leaving the origin C in q_i space, which exhibit no V-variation.

Closely associated with the root structure of a V-surface is the V-variation on a closed loop in coordinate space enclosing the origin C. Taking $\tfrac{1}{2} V_{ij}^{\prime C} q_i q_j$ as a definite quadratic form in all circumstances, we can employ

$$\tfrac{1}{2} V_{ij}^{\prime C} q_i q_j = \text{constant} \tag{6.11}$$

as the closed loop. Stationary values of V on the loop are then given by stationary values of the function

$$V = V(q_i, \Lambda^C, 0) + \tfrac{1}{2}\lambda V_{ij}^{\prime C} q_i q_j$$

where λ is a Lagrange multiplier. But this is exactly the same problem as is posed

134

by the post-buckling of the perfect system, except that in the latter case λ refers to an incremental loading parameter. So, classification based on the closed-loop principle is identical to that based on the paths of the perfect system.

For the two-fold cubic form of equation (6.10) above we thus arrive at the three essentially different situations of Figure 6.15. The elliptic umbilic, at the top, exhibits three roots shown as heavy lines, the light lines depicting contours at non-zero values of V. Stationary values of V are found wherever the ellipse of equation (6.11) touches a contour line, and here one such point arises in each of the six domains between the root lines, as shown. This corresponds to three post-buckling equilibrium paths, and generalizing from the semi-symmetric termi-nology we call such a situation an *anticlinal* point of bifurcation.

The hyperbolic umbilic exhibits just a single root, and the contours have the form at the bottom of Figure 6.15. Here, depending on the aspect ratio and orientation of the ellipse, we may have two or six stationary points as shown. We

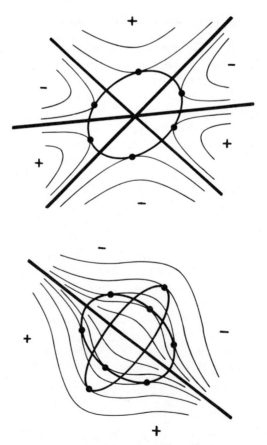

Figure 6.15 Classification of two-fold branching by the cubic form of the potential function at the critical state C

thus have either one or three post-buckling paths giving, by an extension of the semi-symmetric terminology, either a *monoclinal* or a *homeoclinal* point of bifurcation, respectively.

We observe that the classification is finer than Thom's, since the hyperbolic umbilic catastrophe gives both the monoclinal and homeoclinal points of bifurcation, and that, as before, this arises because we have an *a priori* given primary control parameter Λ associated with $V_{ij}^{\prime c} q_i q_j$.

6.5 ROUTES THROUGH THE UMBILIC CATASTROPHES

The notion of 'routes through catastrophes' proved useful in Chapter 4 in distinguishing between different manifestations of the same cuspoid. We now

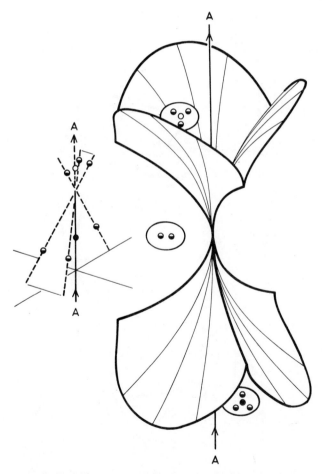

Figure 6.16 Route in control space through the elliptic umbilic catastrophe giving the anticlinal point of bifurcation

examine the umbilics in the same way, viewing a general form of the failure locus in three-dimensional control space, and identifying alternative routes taken by the Λ-axis through each phenomenon. The elliptic umbilic is found to appear as the anticlinal point of bifurcation, and the hyperbolic umbilic takes the three forms of monoclinal, homoclinal, and hilltop branching. These are not the only forms that can arise, but they are the most likely.

The elliptic umbilic

The structurally stable unfolded form of the elliptic umbilic catastrophe is shown in Figure 6.16, on the right. Here a solid circle denotes an energy minimum (stable state), a half-solid circle denotes a saddle point (unstable with one degree of

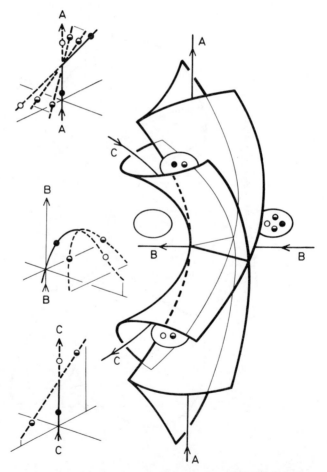

Figure 6.17 Routes in control space through the hyperbolic umbilic catastrophe giving the homoclinal and monoclinal points of bifurcation and the hilltop branching point

instability), and an open circle denotes a maximum (unstable with two degrees of instability).

The control space is divided by the failure surface into three regions, only one of which contains a stable state. A physically realizable control route must therefore enter from this region, and the path A gives us the anticlinal point of bifurcation illustrated on the left. Clearly, if the route A just misses the umbilic point itself we can obtain a perturbed version of anticlinal branching associated with a non-zero value of splitting parameter σ, as shown in Figure 6.8 and at the right of Figure 6.13.

The hyperbolic umbilic

The unfolded form of the hyperbolic umbilic catastrophe in three-dimensional control space is shown on the right of Figure 6.17; this is topologically equivalent to the plots of Figures 6.11 and 6.12. Here the control space is divided by the surfaces into four regions with the equilibrium solutions shown, one region having no solution at all.

Here there is a much more varied choice of route through the singularity than for the elliptic umbilic, and three possibilities are shown. Route A gives the homeoclinal point of bifurcation (top-left), route C gives the monoclinal point of bifurcation (bottom-left), and route B gives the hilltop branching point (middle-left). The last of these has, up to now, been omitted from our discussion since it exhibits limiting behaviour on the fundamental path; we treat it briefly in the following subsection.

As with the fold and the cusp, the introduction of an extra dimension of control can give rise to new variations of a known geometry. One such example, arising in the rotationally symmetric buckling of a complete spherical shell under uniform external pressure (see Chapter 8), is shown in Figure 6.18.[1,13] Here the bilinear corner unique to the hyperbolic umbilic has been stretched out into a third dimension. This represents a structurally *unstable* three-dimensional section, through a four-dimensional control space exhibiting a string of hyperbolic umbilic points.

The hilltop branching point

As mentioned above, the hyperbolic umbilic catastrophe appears in a third bifurcational manifestation as the hilltop branching point. This we have largely ignored until now, since the fundamental equilibrium path of the perfect system exhibits limiting behaviour; it is thus outside the scope of the next chapter, demanding a quite different analytical approach.[18] It is, however, of some practical significance, appearing in the response of a tied arch, for example. Here we shall use the mechanical instability of a stressed atomic crystal lattice[100] to illustrate the phenomenon, following the work of Thompson and Shorrock.[2,20,21]

We consider a close-packed crystal with Lennard–Jones interatomic poten-

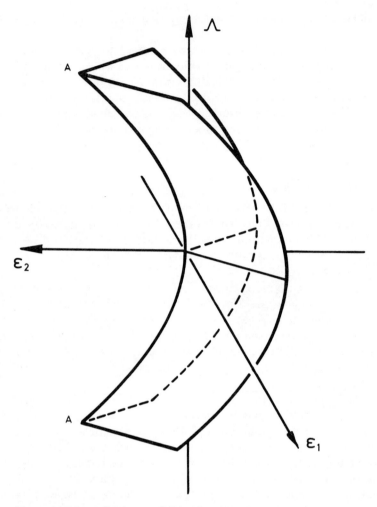

Figure 6.18 Imperfection-sensitivity for the spherical shell. The load and conventional geometric imperfections in the buckling modes fail in this case to fully unfold the hyperbolic umbilic

tials, subjected to a uniaxial tensile stress σ_{11}. This gives rise to a direct strain ε_{11}, and we can show that the response of the system is as given at the top of Figure 6.19. Here we have a primary (fundamental) equilibrium path exhibiting limiting behaviour, but below the limit point there is a symmetry-breaking unstable-symmetric bifurcation into a shear mode ε_{12}. The presence of a shearing stress σ_{12} acts as an imperfection, rounding off the bifurcating paths. Clearly, we would have a cusp in control space $\sigma_{11} - \sigma_{12}$ for the local failure locus.

If a second direct stress σ_{22} is added, the form of the primary path is unaffected, but the bifurcation is moved along the path thereby affecting its stability. We can evaluate the magnitude of this (tensile) stress, σ_{22}^{C}, that is required to shift the

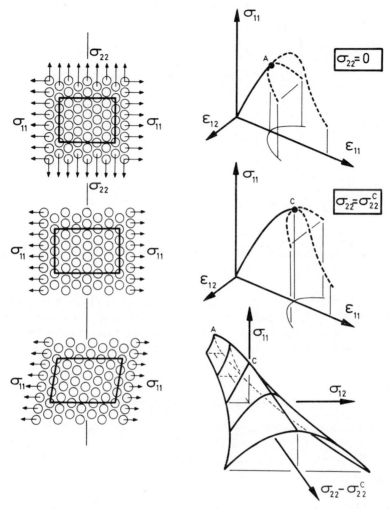

Figure 6.19 Mechanical instablility of a stressed crystal lattice giving a hilltop branching point

bifurcation up to the limit point to generate a hilltop branch, as shown in the middle diagram of Figure 6.19. Here the two-thirds power law cusp is replaced by a bilinear failure locus.

Unfolding this singularity[21] we must obtain a full three-dimensional failure locus of the form shown at bottom-right of Figure 6.19; here all complementary path behaviour is omitted. We identify this once again as the hyperbolic umbilic catastrophe.

A second example of the hilltop branch is provided by the elastically tied arch of Figure 6.20. For a shallow arch, failure is associated with a limit point as shown, while for a deep arch we have an unstable-symmetric bifurcation lying

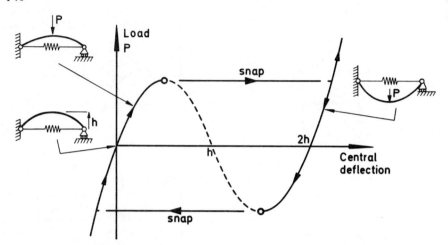

Figure 6.20 The symmetric response of a tied arch. A symmetry-breaking bifurcation on this path allows hill-top branching

below the limit point, as seen in the pinned arch analysis of Chapter 5. Thus the paths take precisely the same form as those of the stressed crystal lattice of Figure 6.19; by adjusting the tension in the spring we can arrange for coincidence, as in the middle diagram. This therefore also generates the form of failure locus shown.

6.6 HIGHER-ORDER TWO-MODE SINGULARITIES

We end the chapter with a brief review of higher-order phenomena. It appears that one of these, the parabolic umbilic catastrophe, is of considerable importance in interactive buckling problems; its bifurcational manifestation is a research topic of current interest.[101,102]

The parabolic umbilic

It seems that the parabolic umbilic catastrophe has received somewhat less attention in the buckling literature than is its due, considering its importance in mode interaction problems.[81,103] Emphasis has traditionally been placed on either semi- or full symmetry rather than on the hybrid situation in which symmetry is present for each contributing mode taken separately but is broken with respect to one of the modes when they both appear simultaneously; we show how this can arise in both stiffened plates and shells, and the axially loaded isotropic cylinder, in Chapter 8.

The solution routines of Chapter 7 have yet to be applied in full to the parabolic umbilic, which is a further step removed in complexity from the elliptic and hyperbolic umbilics. It requires four control parameters for structural stability—the load, two imperfections, and a splitting parameter—and after the elimination of passive coordinates is described by the potential function

$$V = \tfrac{1}{24}V^C_{2222}u^4_2 + \tfrac{1}{2}V^C_{112}u^2_1u_2 + \tfrac{1}{2}(\Lambda - \Lambda^C)(V'^C_{11}u^2_1 + V'^C_{22}u^2_2) + \tfrac{1}{2}\sigma V'^C_{11}u^2_1$$
$$+ V^{1C}_1 \varepsilon^1 u_1 + V^{2C}_2 \varepsilon^2 u_2 + \text{higher-order terms}$$

This is very similar in form to the potential function (6.7) for the hyperbolic and elliptic umbilics, the only difference being that the cubic V^C_{222} term is replaced by the quartic V^C_{2222}. We see that we now have complete symmetry in each of the modes separately (setting one of the u_i to zero leaves a potential function symmetric in the other), but the cross-term V^C_{112} breaks the symmetry with respect to u_2 when both coordinates appear.

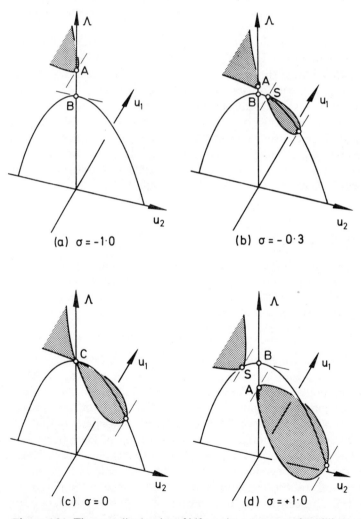

(a) $\sigma = -1\cdot0$

(b) $\sigma = -0\cdot3$

(c) $\sigma = 0$

(d) $\sigma = +1\cdot0$

Figure 6.21 The paraclinal point of bifurcation: patterns of equilibria for a perfect system exhibiting a parabolic umbilic catastrophe

Under a bifurcational formalism the parabolic umbilic appears as the transition between anticlinal and homeoclinal branching, a feature seen for the guyed cantilever in Figure 6.2; it exhibits two falling coupled post-buckling equilibrium paths at C, as shown in Figures 6.8 and 6.9 for the anticlinal and homeoclinal cases, respectively, but has an uncoupled post-buckling path of zero slope. The V^C_{2222} term adds curvature to this path and so the parabolic umbilic, like the cusp, takes both standard and dual forms, depending on whether the curvature is positive or negative; Figure 6.21 illustrates the dual form, exhibiting the distinctive loop of coupled equilibrium states.[101,102] We use the term *paraclinal point of bifurcation* for the bifurcational manifestation of both forms, but shall refrain from attaching 'stable' or 'unstable' to distinguish between them; this is because, unlike the distinct symmetric bifurcations, the critical state C is unstable in both circumstances.

The imperfection-sensitivity plot for the paraclinal point of bifurcation should exhibit some interesting features, as discussed by Hui and Hansen;[104] it will, for example, unlike the earlier cases of semi-symmetric branching, suffer changes in topological form with variation in splitting parameter σ, since the three-dimensional view is structurally unstable. Schematic drawings of the unfolding in control space are given in most catastrophe theory texts,[3–5] following some work by Jänich. This is usually shown as a series of two-dimensional views in imperfection space ε^i, indicating the topological changes as we move through the 'load' space Λ–σ. The information is a little difficult to translate into the three-dimensional views at constant σ that we would like here.

The symbolic umbilic

The symbolic umbilic catastrophe, or E_6 in the language of Arnold for the phenomena outside Thom's list,[105] is yet more complex, requiring five control parameters for structural stability. After the elimination of passive coordinates, it is described by the potential function

$$V = \tfrac{1}{24} V^C_{2222} u_2^4 + \tfrac{1}{6} V^C_{111} u_1^3 + \text{higher-order terms}$$

at the compound critical state C, with none of the control parameters included. For the propped cantilever model of this chapter, it arises at $\alpha = 90$ degrees, as shown in Figure 6.2. A study extending beyond initial slopes has revealed the four post-buckling paths of Figure 6.22;[86] we see that two of the paths have the same initial orientation in load-coordinate space, but subsequently split.

For the symbolic umbilic, the cross V^C_{112} term of the parabolic umbilic is replaced by the cubic term V^C_{111}, which relates to just the u_1 mode alone. It seems unlikely that symmetry considerations could lead a practical structure to display this form, and we thus feel it to be of less interest than the parabolic umbilic. Nevertheless, it does occupy an important place in the substrata of the double cusp, and appears in the sweep through the umbilic bracelet of the guyed cantilever. Its unfolding is explored in considerable depth by Callahan.[106]

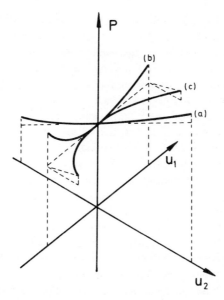

Figure 6.22 Post-buckling paths of the symbolic umbilic catastrophe at complete coincidence of the contributing bifurcations

The double cusp

Lastly, we come to the double cusp catastrophes, or X_9 in the language of Arnold.[105] These exhibit no cubic variation of V at the two-fold compound critical state C. After the elimination of passive coordinates we thus have

$$V = \tfrac{1}{24}V^C_{1111}u_1^4 + \tfrac{1}{6}V^C_{1112}u_1^3u_2 + \tfrac{1}{4}V^C_{1122}u_1^2u_2^2 + \tfrac{1}{6}V^C_{1222}u_1u_2^3$$
$$+ \tfrac{1}{24}V^C_{2222}u_2^4 + \text{higher-order terms}$$

where, as with the symbolic umbilic representation, we have suppressed the effect of all the control parameters.

In Chapter 8 we explore symmetry considerations which lead to the double cusp, and find that symmetry in the combination of both modes is enough to guarantee that all the relevant cubics vanish. If moreover, the potential function is symmetric in each mode in the presence of the other—a stronger symmetry condition—then all terms of odd power in the coordinates vanish and we are left with

$$V = \tfrac{1}{24}V^C_{1111}u_1^4 + \tfrac{1}{4}V^C_{1122}u_1^2u_2^2 + \tfrac{1}{24}V^C_{2222}u_2^4 + \text{higher-order terms}$$

with the control parameters again suppressed.

The double cusp is an important phenomenon, not only for its symmetry properties but because it supplies the umbilics with a *compact geometry* in which to appear; this is discussed in Zeeman's definitive article.[4] A compact geometry is common only to those phenomena with a stable most-critical state C, just the cusp and the butterfly from Thom's list, for example; it implies that the system is somehow 'stable in the large',[1] even though locally it may behave in a quite unstable fashion. Thus, a fold catastrophe seen in the context of a stable cusp describes not just the process of buckling but the end-result, the final deformed state. Similarly here, the global response of a system displaying an umbilic catastrophe is inextricably bound up with the geometry of the double cusp.

Our treatment here of the double cusp is only very rudimentary. The phenomenon is extremely complex, taking four different forms and requiring eight unfolding parameters for structural stability. Much work has been done, both from a bifurcational point of view and following Zeeman.[4] The post-buckling response has been looked at in detail by Supple[91] for both complete and near-coincidence of the contributing bifurcations. The full unfolding is discussed by Magnus and Poston,[5,107] with particular reference to a plate buckling problem described by the von Karman equations.

There is no doubt that the double cusp geometry is important to many plate and shell problems, both stiffened and unstiffened. The following subsection concerns a simple, well-known system, the Augusti model, which exhibits one of the double cusp forms; this contains some hint of the underlying complexity. We note also the clear significance of the double cusp in the evolution of rotating planetary masses.[2]

The Augusti model

This simple two-degree-of-freedom buckling model is due initially to Augusti,[108] later being analysed in relation to an optimization scheme by Thompson and

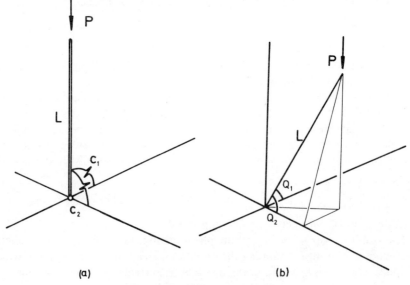

(a)   (b)

Figure 6.23 The Augusti model

Supple.[1,109] We have a light inextensional rod, pinned at its base, and carrying a dead vertical load P, as shown in Figure 6.23. The rod is restrained by two rotational springs of stiffness c_1 and c_2 tied to orthogonal axes as shown; the plane of action of each spring thus rotates as the other spring deforms. Deflections are specified by the two angles spanned by the springs, Q_1 and Q_2, and we introduce the incremental generalized coordinates

$$u_1 = \pi/2 - Q_1, \quad u_2 = \pi/2 - Q_2$$

Imperfections are introduced by specifying initial values of the coordinates, u_1^0 and u_2^0 with the springs unstressed.

To introduce the context of an optimization scheme we suppose that there is a constant amount of spring stiffness to be shared between the springs. The optimum load-carrying capacity is obtained when the springs are of equal stiffness, but here we have complete coincidence of the two contributing bifurcations; each of these corresponds to rotation against one of the springs with the other remaining undeformed, and is of stable-symmetric type when considered in isolation.

The perfect equilibrium paths for this system are given in Figure 6.24. Two cases are shown, first, when the two critical points have a finite separation and second, when they are completely coincident. We see that coupling between the two symmetric and upwards-curving bifurcations gives rise to downwards-curving equilibrium paths at secondary bifurcations, and these are drawn into the primary critical point at coalescence. So, as with the parabolic umbilic catastrophe, two distinct stable forms of branching can coincide to generate imperfection-sensitivity, and we note that this imperfection-sensitivity is mar-

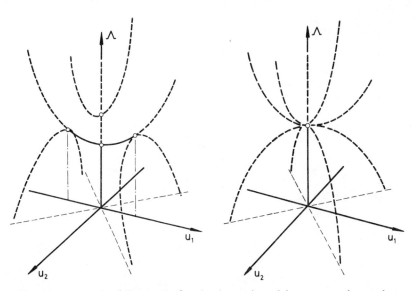

Figure 6.24 Post-buckling paths for the Augusti model at near and complete coincidence of the contributing bifurcations

kedly more severe at complete coincidence than with a finite separation. This acts to erode the apparent optimum of the perfect system, so that with an imperfection in each mode of five degrees the original sharp optimum is severely flattened.[1,109]

The analysis outlined above is clearly incomplete in the light of the complexity of the double cusp topology. Nevertheless, it does seem likely that it embraces the form of imperfection with the most severe possible effect on the response of the system.

7

Comprehensive Bifurcation Analysis

The purpose of this chapter is to provide a systematic and comprehensive method of analysing bifurcations, embracing both distinct and compound branching phenomena. The analysis specifically excludes limit points—a sliding coordinate transformation central to the scheme would be invalid for them—and so the limit point itself and the hilltop branching point cannot be treated by this general method. Otherwise all the phenomena discussed in the early chapters can be handled by the scheme, which can be adapted to suit post-buckling, imperfection-sensitivity, and secondary bifurcation studies, with the contributing bifurcation points either close or completely coincident. The treatment closely follows a recent contribution to the *Philosophical Transactions of the Royal Society*.[103]

We start with our most general system, and assume that it can be described by the gradient potential function

$$V = V(Q_i, \Lambda^j) \tag{7.1}$$

where Q_i is a set of n generalized coordinates and Λ^j a set of h control parameters. We suppose that V is subject to the two axioms given in Chapter 1. The first step is to eliminate $n - m$ *passive coordinates*. During this process the control parameters are largely left unspecified, although we are obliged to identify one with special properties as a distinctive *bifurcational parameter*. Problems of near-coincidence are assumed to be overcome by specific Λ^js which draw the bifurcations together to form an m-fold point of bifurcation. A comprehensive branching analysis is then developed, in terms of just the m active coordinates and the h control parameters. Here a classification of the types of Λ^j parameter is necessary, and the concepts of *generalized imperfection* and *generalized loading parameter* are introduced.

Both parts of the analysis are developed using a general *non-diagonalized* format, and the scheme is thus particularly suited to numerical treatments employing finite elements. Furthermore, coefficients of Taylor expansions are left in general form, rather than being replaced by unity or some other convenient number as is common in more theoretical studies; the analysis can thus be directly and quantitatively applied to an appropriate potential function without preliminary manipulation. The scheme is specifically illustrated for semi-symmetric branching.

7.1 BIFURCATIONAL FORMALISM

Consider a system governed by a gradient potential function $V(Q_i, \Lambda^j)$. We take it as axiomatic first, that a stationary value of V with respect to the generalized coordinates Q_i is necessary and sufficient for the equilibrium of the system, and second, that a complete relative minimum of V with respect to the Q_i is necessary and sufficient for the stability of an equilibrium state. These concepts imply some measure of time, but we assume that the associated response of the system is instantaneous, in the (evolutionary) time-scale of the control parameters. We therefore have the *n equilibrium equations*, $V_i = 0$, and *critical* equilibrium is further defined by the second variation V_{ij} becoming singular. Here, as elsewhere, a subscript denotes partial differentiation with respect to the corresponding generalized coordinate.

Next we assume that one of the controls is special, in the sense that its associated evolution takes place instantaneously in the time-scale of the rest.[6,46] This model has its origins in the general theory of elastic structures, loads and imperfections being traditionally separated both theoretically and experimentally, but clearly has a wider scope for application, particularly in other branches of the physical sciences. We distinguish the special control by the lack of a superscript and designate it simply as Λ, when the occasion demands that it be separately identified.

The fundamental path

In any stability analysis of practical worth it must be assumed that a critical state is reached via some equilibrium path, presumably initially stable. It is a considerable analytical advantage to measure variations from this fundamental state with a sliding set of n incremental coordinates q_i, and transform to a new potential function $W(q_i, \Lambda^j)$. This we shall now do, finding, not surprisingly, that W is a much more convenient analytical tool than V.

But we must start cautiously, and ensure that q_i is a valid set of generalized coordinates in the region of interest. Clearly, for the limit point this would not be the case, for here states exist which could never be expressed by the q_i, those with $\Lambda > \Lambda^c$ in Figure 3.3, for instance. It is for this very practical reason that we insist that the system experiences a bifurcation in the following way.

We identify a single equilibrium path, traced by varying Λ with the remainder of the Λ^j held constant, which always, in the region of interest, has a component in the Λ-direction in $Q_i-\Lambda$ space. This specifically excludes limit points, along with some higher-order phenomena such as the hilltop branch, and the path is thus single-valued with respect to Λ. We refer to this path as the *primary* or *fundamental* path.

In structural mechanics formulations a fundamental path of this type is often exhibited by the *perfect* system, or that envisaged by the designer. It frequently represents some simple, perhaps trivial, solution, a result of underlying symmetries. *Imperfect* systems are generated at different but constant values of others of the Λ^j, the imperfection parameters. Limit points are not excluded from the paths of imperfect systems.

It is well known that, on varying a single control, there is a unique equilibrium path through a non-critical equilibrium state.[1] Moreover, the basic theorems of elastic stability discussed in Section 3.3 show that stability of the fundamental path can, in the absence of limiting behaviour, only be lost at an intersection with a second path, a point of bifurcation. We see later that, under such circumstances, the potential function carries special properties with respect to the single control Λ, which we henceforth refer to as the *bifurcational parameter*.

Sliding coordinates and associated transformation

Formally, the single-valued fundamental path is written

$$Q_i = Q_i^F(\Lambda) \tag{7.2}$$

and we assume it to be known. It gives a path F in Q_i–Λ space, which varies monotonically with Λ in the region of interest, and which intersects other as yet unknown *secondary* or *post-buckling* paths at points of bifurcation.

We introduce a localized set of n incremental sliding coordinates q_i defined by

$$Q_i = Q_i^F(\Lambda) + q_i \tag{7.3}$$

implying a one-to-one correspondence between the Q_i and the q_i; this is clearly only true in the absence of limit points on F. With varying Λ the origin of the new coordinate frame slides along the fundamental path, spanning the full range of interest of Q_i–Λ space; again it is only by denying limiting behaviour that this can be ensured.

We write a new potential function, in terms of the q_i,

$$W(q_i, \Lambda^j) \equiv V[Q_i^F(\Lambda) + q_i, \Lambda^j] \tag{7.4}$$

This merely amounts to substituting equation (7.3) into V; any constants generated can be ignored, since we are interested only in variations of V or W, never absolute values. However, the manoeuvre may destroy a linearity in Λ, as discussed in our earlier book:[1] if V is linear in Λ, W may not be.

The equilibrium and stability condition, expressed by our two axioms, pass over unchanged to the new W-function, which has the properties

$$W_i^F = W_i'^F = W_i''^F = W_i'''^F = \cdots = 0 \tag{7.5}$$

where subscripts denote partial differentiation with respect to the q_i and a prime denotes partial differentiation with respect to Λ. These arise because, in mapping from Q_i–Λ space to q_i–Λ space, F is moved onto the Λ-axis, and a Taylor expansion of W_i in terms of Λ must thus result in $W_i = 0$ everywhere.

The derivative W_i' is of most interest here. Its vanishing at a point of bifurcation is intimately connected with the fact that here, as opposed to a limit point, a generalized load does no first-order work as an elastic structure moves through its buckling displacement. So Λ must then act initially on a quadratic form of the generalized coordinates in a Taylor expansion of the potential function about the critical point, instead of a linear form, as is the case with the limit point.[1]

Active and passive coordinates

Stability of equilibrium states on the fundamental path is governed, in the first instance, by the quadratic form $\frac{1}{2}W^F_{ij}q_iq_j$, and we assume that the path is initially stable, so that for low Λ, W^F_{ij} is positive–definite. With increasing Λ we shall suppose that we encounter an *m-fold compound point of bifurcation* C, where $\Lambda = \Lambda^C$ and W^C_{ij} is singular and of rank $n - m$(co-rank m). We assume that no other bifurcations arise on F in the region of interest, so problems of near-coincidence are to be treated with the *a priori* introduction of specific Λ^j to draw the bifurcations together.

We next split the n generalized coordinates into two distinct subsets, the active and passive coordinates. Thus $n - m$ are selected as the *passive* coordinates and we adopt Greek suffices for their use, writing them as q_α, while the remaining m become the *active* coordinates and we henceforth reserve Latin subscripts for their use. The only restricting condition on this segregation is that the submatrix related to just the passive coordinates must be non-singular, and so

$$|W^C_{\alpha\beta}| \neq 0 \qquad (7.6)$$

We know from the definition of the rank of a matrix that a valid segregation can always be found, but we note that it is not necessarily unique; a different choice of passive coordinates could lead to a different but equally valid analysis. However, the situation is completely clear-cut if the quadratic form is diagonalized, for then the active coordinates sensibly must be the amplitudes of the critical modes.

7.2 ELIMINATION OF PASSIVE COORDINATES

We next demonstrate how the $n - m$ passive coordinates can be eliminated from the analysis by the systematic use of an intrinsic perturbation scheme.[1,79] This general method has proved most useful and versatile in stability studies,[93,110-12] and is to be the major analytical tool of this chapter. The scheme for eliminating the passive coordinates serves as an introduction to the general underlying philosophy.

The procedure can be described as follows. We start by assuming, in implicit parametric form, a result we are trying to obtain. This is then substituted into the appropriate non-linear equations—those which, if they could be solved, would give the required result. The equations become identities—we effectively constrain the system to states which satisfy them—and can be repeatedly differentiated term by term to generate an ordered series of equations, often sequentially linear. Particular solutions to these perturbation equations can then be used to construct the assumed form as a power series.

The implicit assumption that non-linear relations can be expressed as Taylor series raises certain mathematical questions which we shall not attempt to answer here; we merely refer to the excellent account of Poston,[113] who discusses the point in the light of the alternative underlying assumptions of analyticity and determinacy. It suffices to say that for all applications considered here, the assumption is adequately justified.

Here we begin by assuming that the passive coordinates can be written as functions of the actives and the control parameters, thus

$$q_\alpha = q_\alpha(q_i, \Lambda^j) \tag{7.7}$$

which are then substituted into the passive equilibrium equations $W_\alpha = 0$ to give the identities

$$W_\alpha[q_i, q_\beta(q_i, \Lambda^j), \Lambda^j] \equiv 0 \tag{7.8}$$

These define an *activity surface* in the full $q_\alpha - q_i - \Lambda^j$ space, which in essence contains an adequate description of the system.[1]

The left-hand side is now just a function of q_i and Λ^j, and can be repeatedly differentiated with respect to these to give the ordered series of perturbation equations. Differentiating once, we have

$$W_{\alpha i} + W_{\alpha\beta} q_{\beta i} = 0 \tag{7.9}$$
$$W_{\alpha\beta} q_\beta^j + W_\alpha^j = 0$$

and a second time,

$$W_{\alpha i j} + W_{\alpha\beta i} q_{\beta j} + W_{\alpha\beta j} q_{\beta i} + W_{\alpha\beta\gamma} q_{\beta i} q_{\gamma j} + W_{\alpha\beta} q_{\beta i j} = 0$$
$$W_{\alpha\beta i} q_\beta^j + W_{\alpha i}^j + W_{\alpha\beta\gamma} q_{\beta i} q_\gamma^j + W_{\alpha\beta}^j q_{\beta i} + W_{\alpha\beta} q_{\beta i}^j = 0 \tag{7.10}$$
$$W_{\alpha\beta\gamma} q_\beta^i q_\gamma^j + W_{\alpha\beta}^i q_\beta^j + W_{\alpha\beta} q_\beta^{ij} + W_{\alpha\beta}^j q_\beta^i + W_\alpha^{ij} = 0 \quad \text{etc.}$$

Here subscripts on W denote partial differentiation (of the original *unconstrained* form, before the substitution of equation (7.7)) with respect to the corresponding coordinate, as do *further* subscripts on q_β; similarly, superscripts denote partial differentiation with respect to the corresponding control parameter. We thus have

$$q_{\beta i} \equiv \frac{\partial q_\beta}{\partial q_i}, \; q_\beta^j \equiv \frac{\partial q_\beta}{\partial \Lambda^j}, \; q_{\beta i}^j \equiv \frac{\partial^2 q_\beta}{\partial q_i \partial \Lambda^j} \quad \text{etc.} \tag{7.11}$$

The tensor summation convention is employed over Greek subscripts, with summation ranging over the $n - m$ passive values.

We now insist that W is written as a Taylor series, expanded about any point on the fundamental path in the region of interest, and the W-derivatives at F are known to as high an order as is necessary. Evaluating the perturbation equations at F then gives the required q_β derivatives. We have the two sets of first-order equations

$$W_{\alpha\beta} q_{\beta i}|^F = - W_{\alpha i}^F$$
$$W_{\alpha\beta} q_\beta^j|^F = - W_\alpha^{jF} \tag{7.12}$$

which are linear, non-singular by equation (7.6), and can be readily inverted to give $q_{\beta i}^F$ and q_{β}^{jF}. The equations are in a standard form, so each of these derivatives can be written directly as the ratio of two determinants, the denominator being the complete determinant of $W_{\alpha\gamma}^F$ and the numerator being the determinant of this same matrix with the particular $W_{\alpha\beta}^F$ elements replaced by $-W_{\alpha i}^F$ (or $-W_{\alpha}^{jF}$). We see that this process must comprise a total of $(m + h)(n - m)$ such calculations and may thus require the aid of a digital computer, to which it would be well suited.

We remember that the special bifurcational parameter Λ is among the $\Lambda^j s$, and has the properties of equation (7.5), so we thus have the particular results arising from the second set of equations

$$q_\beta'^F = 0 \tag{7.13}$$

We note that the prime is again used, replacing one of the superscripts, when differentiation is specifically carried out with respect to Λ.

The second-order equations can now be written in the form

$$
\begin{aligned}
W_{\alpha\beta}q_{\beta ij}|^F &= -(W_{\alpha ij} + W_{\alpha\beta i}q_{\beta j} + W_{\alpha\beta j}q_{\beta i} + W_{\alpha\beta\gamma}q_{\beta i}q_{\gamma j})|^F \\
W_{\alpha\beta}q_{\beta i}^j|^F &= -(W_{\alpha i}^j + W_{\alpha\beta i}q_\beta^j + W_{\alpha\beta}^j q_{\beta i} + W_{\alpha\beta\gamma}q_{\beta i}q_\gamma^j)|^F \\
W_{\alpha\beta}q_\beta^{ij}|^F &= -(W_\alpha^{ij} + W_{\alpha\beta}^i q_\beta^j + W_{\alpha\beta}^j q_\beta^i + W_{\alpha\beta\gamma}q_\beta^i q_\gamma^j)|^F
\end{aligned} \tag{7.14}
$$

We see that the right-hand side of each of these is known once the first-order derivatives are obtained, and the equations can thus be readily solved by the same method as above for the derivatives $q_{\beta ij}^F$, $q_{\beta i}^{jF}$, and q_β^{ijF}. The scheme can then be continued to the next-order equations if necessary, and carried on in sequentially linear fashion. One general result, which can readily be proved by induction, is that

$$q_\beta'^F = q_\beta''^F = q_\beta'''^F = \cdots = 0 \tag{7.15}$$

arising from the special properties of the bifurcational control parameter Λ.

We can now write Taylor expansions of $q_\alpha(q_i, \Lambda^j)$ to as high an order as we please, and this leads to a new potential function with only m degrees of freedom, defined by the identity

$$\mathscr{W}(q_i, \Lambda^j) \equiv W[q_i, q_\alpha(q_i, \Lambda^j), \Lambda^j] \tag{7.16}$$

We obtain derivatives of the new function by direct differentiation,

$$
\begin{aligned}
\mathscr{W}_i &= W_i + W_\alpha q_{\alpha i} \\
\mathscr{W}_{ij} &= W_{ij} + W_{\alpha i}q_{\alpha j} + W_{\alpha j}q_{\alpha i} + W_{\alpha\beta}q_{\alpha i}q_{\beta j} + W_\alpha q_{\alpha ij} \\
\mathscr{W}_i^j &= W_i^j + W_{\alpha i}q_\alpha^j + W_\alpha^j q_{\alpha i} + W_{\alpha\beta}q_{\alpha i}q_\beta^j + W_\alpha q_{\alpha i}^j \quad \text{etc.}
\end{aligned} \tag{7.17}
$$

and evaluation on the fundamental path now gives, after a little algebra, the results of interest,

$$
\begin{aligned}
\mathscr{W}_i^F &= 0 \\
\mathscr{W}_{ij}^F &= W_{ij} + W_{\alpha i}q_{\alpha j}|^F \\
\mathscr{W}_i^{jF} &= W_i^j + W_{\alpha i}q_\alpha^j|^F
\end{aligned} \tag{7.18}
$$

$$\mathscr{W}_{ijk}^F = W_{ijk} + W_{\alpha jk}q_{\alpha i} + W_{\alpha ik}q_{\alpha j} + W_{\alpha ij}q_{\alpha k} + W_{\alpha\beta i}q_{\alpha j}q_{\beta k}$$
$$+ W_{\alpha\beta j}q_{\alpha i}q_{\beta k} + W_{\alpha\beta k}q_{\beta i}q_{\alpha j} + W_{\alpha\beta\gamma}q_{\alpha i}q_{\beta j}q_{\gamma k}|^F$$

$$\mathscr{W}_{ij}^{kF} = W_{ij}^k + W_{\alpha j}^k q_{\alpha i} + W_{\alpha i}^k q_{\alpha j} + W_{\alpha ij}q_\alpha^k + W_{\alpha\beta i}q_{\alpha j}q_\beta^k + W_{\alpha\beta j}q_{\alpha i}q_\beta^k$$
$$+ W_{\alpha\beta}^k q_{\alpha i}q_{\beta j} + W_{\alpha\beta\gamma}q_{\alpha i}q_{\beta j}q_\gamma^k|^F$$

$$\mathscr{W}_{ijkl}^F = W_{ijkl} + W_{\alpha jkl}q_{\alpha i} + W_{\alpha ikl}q_{\alpha j} + W_{\alpha ijl}q_{\alpha k} + W_{\alpha ijk}q_{\alpha l}$$
$$+ W_{\alpha\beta kl}q_{\alpha i}q_{\beta j} + W_{\alpha\beta jl}q_{\alpha i}q_{\beta k} + W_{\alpha\beta jk}q_{\alpha i}q_{\beta l} + W_{\alpha\beta il}q_{\alpha j}q_{\beta k}$$
$$+ W_{\alpha\beta ik}q_{\alpha j}q_{\beta l} + W_{\alpha\beta ij}q_{\alpha k}q_{\beta l} + W_{\alpha\beta\gamma i}q_{\alpha j}q_{\beta k}q_{\gamma l} + W_{\alpha\beta\gamma j}q_{\alpha i}q_{\beta k}q_{\gamma l}$$
$$+ W_{\alpha\beta\gamma k}q_{\alpha i}q_{\beta j}q_{\gamma l} + W_{\alpha\beta\gamma l}q_{\alpha i}q_{\beta j}q_{\gamma k} + W_{\alpha\beta\gamma\delta}q_{\alpha i}q_{\beta j}q_{\gamma k}q_{\delta l}$$
$$- W_{\alpha\beta}q_{\alpha ij}q_{\beta kl} - W_{\alpha\beta}q_{\alpha ik}q_{\beta jl} - W_{\alpha\beta}q_{\alpha il}q_{\beta jk}|^F \quad \text{etc.}$$

which can be used to construct a Taylor expansion of $\mathscr{W}$. The equilibrium and stability conditions (expressed by our axioms) pass over unchanged to the new function,[1] so from now on we can operate exclusively and with confidence in terms of $\mathscr{W}$. We have in particular the special results

$$\mathscr{W}_i^F = \mathscr{W}_i^{'F} = \mathscr{W}_i^{''F} = \cdots = 0$$
$$\mathscr{W}_{ij}^{'F} = W_{ij}' + W_{\alpha i}'q_{\alpha j} + W_{\alpha j}'q_{\alpha i} + W_{\alpha\beta}'q_{\alpha i}q_{\beta j}|^F \qquad (7.19)$$

from the properties of the bifurcational parameter Λ, and at the m-fold critical point itself,

$$\mathscr{W}_{ij}^C = 0 \quad \text{(all } i, j) \qquad (7.20)$$

so that $\mathscr{W}_{ij}^C$ is null.[1]

This completes the general perturbation method for the elimination of $n - m$ passive coordinates at an m-fold point of bifurcation, which, as we discuss earlier, corresponds to the splitting lemma of mathematical texts; rigorous justification of the process lies in the implicit function theorem.[5] The scheme makes no resort to diagonalization of the potential function, and thus is particularly suited to numerical computer analysis, arising from a finite-element or finite-difference formulation. We note that, although we are now able to cast a problem in an m-dimensional form, the passive coordinates are not simply neglected, and any contaminating effect that they may have on the buckling modes is automatically taken into account.

7.3 PERTURBATION ANALYSIS

We present in this section a comprehensive perturbation approach to the analysis of m-fold bifurcation. The general treatment allows us to set up the appropriate first-order equations over a very broad spectrum of problems, including post-buckling, imperfection-sensitivity, loci of secondary bifurcations, and even the locus of an m-fold point itself. Complete solution to the equations, to give asymptotically exact results for the problem in hand, is often difficult, and we demonstrate here two particular aids to solution in the twin concepts of *generalized imperfection* and *generalized load*.

The complete process is illustrated in the comprehensive analysis of semi-symmetric branching, giving rise to the quantitative plots of the previous chapter. A numerical search procedure is described to determine the imperfection-sensitivity when the two contributing bifurcations of the perfect system exhibit a finite separation on the fundamental path F.

Having successfully eliminated the passive coordinates we shall operate exclusively on the m-degree-of-freedom potential function

$$\mathscr{W} = \mathscr{W}(q_i, \Lambda^j) \tag{7.21}$$

with the properties described at the end of the previous section, q_i denoting the active coordinates. The set of control parameters Λ^j, of course, includes the distinctive bifurcational parameter Λ but the composition of the remainder depends crucially on the problem in hand; thus, for example, m imperfection parameters can be identified among the Λ^j, but these are set to zero to investigate the equilibrium behaviour of the perfect system.

The governing equations to be used depend precisely on the information required. The full set of equilibrium equations $\mathscr{W}_i = 0$ will always appear. A general point of critical equilibrium is further defined by the local eigenvalue equation $\mathscr{W}_{ij}x_j = 0$, where x_j denotes a local eigenvector.[93] This takes the specific form $\mathscr{W}_{ij} = 0$ if the locus of an m-fold critical point is sought.[92] Secondary bifurcations are located from amongst all critical states by the additional equation $\mathscr{W}'_i x_i = 0$.[93] A perturbation approach with a single independent variable is to be employed, so all variables, including the local eigenvector x_k, are written in the parametric form

$$q_i = q_i(s), \ \Lambda^j = \Lambda^j(s), \ x_k = x_k(s) \tag{7.22}$$

where s is the single *perturbation parameter*, as yet unidentified; it may be left in this general form or specified further, according to the demands of the analysis. Perturbation equations are to be derived initially in the most general context, but in any specific instance certain terms may not be present, as we shall see in the illustration of semi-symmetric branching.

Equilibrium equations

Substituting the above parametric forms into the equilibrium equations $\mathscr{W}_i = 0$ we obtain the *equilibrium identity*

$$\mathscr{W}_i[q_j(s), \Lambda^k(s)] \equiv 0 \tag{7.23}$$

which is differentiated once with respect to s to give

$$\mathscr{W}_{ij}q_j^{(1)} + \mathscr{W}_i^j\Lambda^{j(1)} = 0 \tag{7.24}$$

Here a superscript in parentheses denotes the number of full differentiations with respect to the perturbation parameter s, and the tensor summation convention is employed, repeated subscripts denoting summation from one to m and repeated superscripts denoting summation from one to h.

On evaluation at the m-fold bifurcation point C the first term vanishes by the criticality condition (7.20), and we are left with

$$\mathscr{W}_i^{\,j}\Lambda^{j(1)}|^C = 0 \tag{7.25}$$

which at first glance seems a set of m equations in the h unknowns $\Lambda^{j(1)C}$. But some of this set of derivatives make no appearance in the equations, since they are multiplied by zero coefficients; the bifurcational parameter derivative $\Lambda^{(1)C}$ would be absent because of conditions (7.19), and we might also expect that the derivative of the splitting parameter of the previous chapter would likewise not be present.

The control parameters that are represented are those which act on a linear form of the active coordinates q_i, in a Taylor expansion of $\mathscr{W}$ about C. These are, of course, the imperfection parameters ε^j, and we shall suppose there to be m in number; if this is too many for a specific problem we can later set some to zero, while a formulation with more than m can always be reduced to this number of *independent* parameters by lumping similar contributions together. This reduces equation (7.25) to m equations in m unknowns,

$$\mathscr{W}_i^{\,j}\varepsilon^{j(1)}|^C = 0 \tag{7.26}$$

and assuming that the imperfections arise in the system in some typical fashion so as to render these equations non-singular and $|\mathscr{W}_i^{\,jC}| \neq 0$, we have the important result

$$\varepsilon^{j(1)} = 0 \tag{7.27}$$

The m imperfection parameters will be assumed, for the sake of simplicity, to be the first of the Λ^j, with j ranging from one to m. The remaining $h - m$ control parameters, which must comprise a non-empty set since it includes the bifurcational parameter Λ, are formally represented by the condition

$$\mathscr{W}_i^{\,jC} = 0 \tag{7.28}$$

for all i and j ranging from $m + 1$ to h.

The segregation of the controls into two groups is of deeper significance than might at first be supposed. We shall find in every perturbation equation derived, that the two remain completely separate. They thus clearly have quite different roles to play. The point is well illustrated in the next equilibrium perturbation equation.

Differentiating the equilibrium identity a second time and evaluating at the point C we obtain

$$\mathscr{W}_{ijk}q_j^{(1)}q_k^{(1)} + 2\mathscr{W}_{ij}^{\,k}q_j^{(1)}\Lambda^{k(1)} + \mathscr{W}_i^{\,j}\varepsilon^{j(2)}|^C = 0 \tag{7.29}$$

the first non-trivial equation of interest. We see that the summation implied by the repeated superscript of the second term ranges just over $k = m + 1$ to h, by virtue of result (7.27), while the third term contains merely ε^j derivatives from conditions (7.28). We have here m non-linear equations in $m + h$ unknowns, and general solution is clearly out of the question; however, they are tractable in certain instances, as we shall see later. The process can be continued to higher order by further differentiation and evaluation, but this is not to be done explicitly here.

Critical state equations

In imperfection-sensitivity studies it is necessary to pinpoint states of critical equilibrium, and this can be done via a local linear eigenvalue equation $\mathscr{W}_{ij}x_j = 0$, x_j representing the local eigenvector.[93] Substituting the parametric forms (7.22) we thus obtain the *critical state identity*

$$\mathscr{W}_{ij}[q_k(s), \Lambda^l(s)]x_j(s) \equiv 0 \qquad (7.30)$$

Differentiation with respect to s and evaluation at C then gives the equations of interest

$$\mathscr{W}_{ijk}x_jq_k^{(1)} + \mathscr{W}_{ij}^k x_j\Lambda^{k(1)}|^C = 0 \qquad (7.31)$$

in general a set of m equations in $m + h$ unknowns. We see that the ε^j derivatives make no appearance here, and the superscript summation of the second term is again over the range $k = m + 1$ to h. The process can be continued by further differentiation and evaluation, but this is not to be given explicitly.

In some studies it may be desirable to trace the locus of an m-fold critical point,[92] and we can do this by replacing the above with its special form for m-fold criticality, $\mathscr{W}_{ij} = 0$. Substituting our parametric forms gives the *m-fold critical state identity*

$$\mathscr{W}_{ij}[q_k(s), \Lambda^l(s)] \equiv 0 \quad (\text{all } i, j) \qquad (7.32)$$

and differentiation and evaluation now gives the m^2 equations in h unknowns

$$\mathscr{W}_{ijk}q_k^{(1)} + \mathscr{W}_{ij}^k\Lambda^{k(1)}|^C = 0 \qquad (7.33)$$

We note that duplications will arise among this set of equations, from the symmetry properties of the $\mathscr{W}$-derivatives. Again the ε^j derivatives make no appearance, superscript summation is over $k = m + 1$ to h, and the scheme could be continued if necessary.

Secondary bifurcation equations

Finally, we may wish specifically to locate secondary bifurcations from amongst all the critical states that can arise. We do this with the introduction, along with equilibrium and critical state equations, of the additional equation $\mathscr{W}_i'x_i = 0$, where x_i denotes the critical local eigenvector, as before. This states algebraically that a generalized load does no first-order work as a system moves through its buckling displacement, and can be rigorously justified by the introduction of a set of incremental coordinates which remain fixed at the critical point of interest.[1] We note that the bifurcational parameter Λ now must play a key role, and the prime appears in the above equation, since the difference between a bifurcation and a limit point depends upon the orientation of the Λ-direction in control space, as we see in Chapter 4.

Substituting the parametric forms (7.22) into this equation we obtain the *secondary bifurcation identity*

$$\mathcal{W}'_i[q_j(s), \Lambda^k(s)]x_i(s) \equiv 0 \tag{7.34}$$

differentiation with respect to s and evaluation at C now giving the single equation in $m + h - 1$ unknowns

$$\mathcal{W}'_{ij}x_iq_j^{(1)} + \mathcal{W}'^j_i x_i\Lambda^{j(1)}|^C = 0 \tag{7.35}$$

The second term must be included in this general formulation. However, ε^j derivatives will be absent as before, derivatives of Λ likewise by conditions (7.19), and similar conditions for the remaining Λ^j derivatives may be such that the complete term vanishes; this is certainly the case for the illustrated semi-symmetric branching analysis. As before, we note that the scheme can be continued if necessary.

7.4 GENERALIZED IMPERFECTIONS

Having shown how to set up the appropriate perturbation equations, we now turn our attention to the question of their solution. For a distinct critical point with $m = 1$ this can usually be easily done,[1] but with $m \neq 1$ the situation becomes a great deal more complicated. However, the analysis can frequently be simplified by systematically taking sections through the control space, thereby reducing the unknowns in a typical set of equations but solving them repeatedly. We now introduce one way in which this can be done, with the concept of a *generalized imperfection*.[93] Here it is assumed that the full imperfection space ε^i is most conveniently scanned with a sweep of an *imperfection ray R* through the full space, making use of a polar rather than a cartesian representation of the imperfections.

Let us therefore consider in isolation this m-dimensional imperfection space ε^i. We introduce a polar coordinate transformation on the imperfection parameters thus,

$$\varepsilon^i = \varepsilon^i(\theta_\alpha, \varepsilon) = G^i(\theta_\alpha)\varepsilon \tag{7.36}$$

where α ranges from 1 to $m - 1$. Here the $G^i(\theta_\alpha)$ are trigonometric functions only; for $m = 2$ they are $G^1 = \sin\theta$, $G^2 = \cos\theta$, while for $m = 3$, $G^1 = \sin\theta_1$, $G^2 = \sin\theta_2\cos\theta_1$, $G^3 = \cos\theta_2\cos\theta_1$, suffice. The parameter ε, which clearly represents progress along a ray emerging from the origin $\varepsilon^i = 0$, is termed the *generalized imperfection*.

Writing now the potential function as $\mathcal{W}(q_i, \Lambda^j, \varepsilon^k)$, so that the set Λ^j is from now on understood specifically to exclude the m imperfection parameters, we can determine the energy level just on the ray R, where $\theta_\alpha = \theta_\alpha^R$, by the transformation

$$\mathcal{W}(q_i, \Lambda^j, \varepsilon) \equiv \mathcal{W}[q_i, \Lambda^j, G^k(\theta_\alpha^R)\varepsilon] \tag{7.37}$$

The derived perturbation equations still hold good for the new potential function of the left-hand side, which we continue to call $\mathcal{W}$, but the number of unknowns is reduced by $m - 1$. Required derivatives can be found by successive differentiation and evaluation at C; those with respect to just q_i and Λ^j remain unchanged, while a dot will be used to denote partial differentiation with respect to the generalized imperfection ε. Thus

$$\dot{\mathcal{W}}_i^C = \mathcal{W}_i^{jC} G^j(\theta_\alpha^R) \quad \text{etc.} \tag{7.38}$$

the repeated superscript implying summation over the range $j = 1$ to m. We see that the use of the same symbol, $\mathcal{W}$, on both sides of the transformation identity need create no confusion, since the syntax will always make it perfectly clear which representation applies. We note finally that the sweep of the ray through the full imperfection space implies considerable repetition, so the use of a computer becomes almost essential in problems of compound bifurcation.

7.5 GENERALIZED LOADS

The generalized imperfection construction is only appropriate when the m imperfection parameters all enter the potential function in the same way, acting on a linear form of the q_i. We now concentrate on the set of Λ^j which act on quadratic terms, which we can suppose includes the bifurcational parameter Λ. When more than one such parameter is to be considered, it is sometimes useful to adopt a similar construction, thereby reducing the number to one but generating a full series of like problems as before.

In a bifurcational problem of structural mechanics, this subset of Λ^j can comprise just structural loading parameters, and so the analysis embraces the multiple-loading situation. The single key parameter of the construction is thus referred to as the *generalized loading parameter*, although it need not carry any such physical significance in a specific problem. The set may include controlled geometric change which act as *splitting parameters* to separate contributing bifurcations on the fundamental path,[87] or possibly other perturbations which might enter the potential function in this special way. The formulation thus corresponds closely to that of Huseyin.[44]

Let us suppose that we have r such parameters λ^j, where $r \leq h - m$, defined by

$$\lambda^j = \Lambda^j - \Lambda^{jC} \tag{7.39}$$

j scanning over the values $m + 1$ to $m + r$. The λ^j are thus defined as incremental measures of the controls away from the critical point C. We see that no attempt has been made to relate the number r to m, unlike the generalized imperfection analysis.

We introduce the polar transformation

$$\lambda^i = \lambda^i(\phi_\alpha, \lambda^*) = H^i(\phi_\alpha)\lambda^* \tag{7.40}$$

where α ranges from 1 to $r - 1$. As with the imperfections, the $H^i(\phi_\alpha)$ are trigonometric functions only, so for $r = 2$ they are simply $H^{m+1} = \sin \phi$, $H^{m+2} =$

$\cos \phi$, etc. The new parameter λ^* represents progress along a ray R emerging from the origin in 'load' space λ^i, and is termed the *generalized loading parameter*.

Writing now the potential function as $\mathscr{W}(q_i, \Lambda^j, \varepsilon, \lambda^k)$, so that Λ^j is here understood to exclude both the m imperfections and the r parameters λ^k and may well be an empty set, we can determine the energy level along the 'load' ray R, where $\phi_\alpha = \phi_\alpha^R$, by the transformation

$$\mathscr{W}(q_i, \Lambda^j, \varepsilon, \lambda^*) \equiv \mathscr{W}[q_i, \Lambda^j, \varepsilon, H^k(\phi_\alpha^R)\lambda^*] \tag{7.41}$$

Again the perturbation equations can be directly written down for the new potential function, which we still call $\mathscr{W}$, but the number of unknowns is reduced by $r - 1$. The required derivatives can be found by successive differentiation and evaluation at C, derivatives with respect to just q_i, Λ^j, or ε remaining unchanged. As asterisk will be used to denote differentiation with respect to λ^*, so we have

$$\mathscr{W}_i^{*C} = \mathscr{W}_i^{jC} H^j(\phi_\alpha^R) = 0 \tag{7.42}$$

from conditions (7.28) and

$$\mathscr{W}_{ij}^{*C} = \mathscr{W}_{ij}^{kC} H^k(\phi_\alpha^R) \tag{7.43}$$

the repeated superscript here, of course, implying summation merely over the range $k = m + 1$ to $m + r$. Again the use of the same symbol, $\mathscr{W}$, on both sides of the transformation identity need create no confusion, since the syntax makes it clear which representation applies.

We note finally that when one of the λ^i is a splitting parameter, as in the following analysis of semi-symmetric branching, the imperfection-sensitivity is most usefully shown with this parameter held constant. To draw such plots a numerical search procedure can be used. This is described during the analysis.

Further aids to solution

The unknowns of a specific problem can be further reduced both by specifying the perturbation parameter s and by normalizing the critical local eigenfunctions x_j. However, both manoeuvres can lead to solutions being missed unless they are sought as special cases. We elaborate on these points in the analysis of semi-symmetric branching.

The concepts of generalized imperfection and generalized load are certainly useful for semi-symmetric analysis, as seen below, and presumably carry an equal significance for other problems dominated by cubic terms of potential energy. However, for more complex phenomena, more subtle transformations may be in order. We cite the example of the parabolic umbilic catastrophe, of great significance in problems of interactive buckling, as seen in Chapter 8. Here, an analysis which takes account of the concept of determinacy[101,102] suggests that a cubic and a quartic energy term carry the same first-order, asymptotic, significance; this is briefly discussed in Section 8.5. It leads to expanded forms of the two imperfections, ε^1 and ε^2, starting initially with terms of order s^5 and s^6, respectively, s being the perturbation parameter. A generalized imperfection measuring progress along a five-sixths law curve in ε^i space here seems more appropriate.

7.6 ILLUSTRATION OF COMPOUND SEMI-SYMMETRIC BUCKLING

To illustrate the perturbation analysis in detail we now present the extensive study of semi-symmetric bifurcation which led to the plots of the previous chapter. We thus suppose that $m = 2$, and that with the imperfections set to zero the potential function is symmetric in one of the contributing critical modes, in the sense that equal and opposite deformations just in that mode give identical energy levels. Odd powers of this amplitude in a Taylor expansion of the potential function at C thus vanish, leaving it diagonalized; we write this as $\mathscr{A}$ rather than $\mathscr{W}$, the active coordinates becoming the amplitudes of the two critical modes, u_i, rather than q_i, to fall into line with an earlier notation.[1]

With symmetry in u_1 we also have $\mathscr{A}^C_{111} = \mathscr{A}^C_{122} = 0$, so introducing the incremental loading parameter $\lambda = \Lambda - \Lambda^C$ and the splitting parameter σ we have the general form

$$\mathscr{A} = \tfrac{1}{6}\mathscr{A}^C_{222}u_2^3 + \tfrac{1}{2}\mathscr{A}^C_{112}u_1^2 u_2 + \tfrac{1}{2}\lambda(\mathscr{A}'^C_{11}u_1^2 + \mathscr{A}'^C_{22}u_2^2) + \tfrac{1}{2}\sigma\mathscr{A}'^C_{11}u_1^2$$
$$+ \mathscr{A}^{1C}_1\varepsilon^1 u_1 + \mathscr{A}^{2C}_2\varepsilon^2 u_2 + \text{higher-order terms} \tag{7.44}$$

To study the simplest possible critical behaviour which can be described by this we must insist that $\mathscr{A}^C_{222}$ and $\mathscr{A}^C_{112}$ both are non-zero. However, we can and do set $\mathscr{A}^{1C}_2 = \mathscr{A}^{2C}_1 = 0$ without loss of generality, so ε^1 and ε^2 are *principal imperfections*, each directly related to a corresponding *principal coordinate* u_1 or u_2.[93] The splitting parameter σ, which when introduced in this way is a direct measure of critical load separation on the Λ-axis, provides for the possibility of analysis with up to four dimensions of control.

The lowest-order equilibrium perturbation equations of interest (7.29) now become

$$\left.\begin{aligned} 2\mathscr{A}_{112}u_1^{(1)}u_2^{(1)} + 2\mathscr{A}'_{11}u_1^{(1)}\Lambda^{(1)} + 2\mathscr{A}'_{11}u_1^{(1)}\sigma^{(1)} + \mathscr{A}^1_1\varepsilon^{1(2)}|^C = 0 \\ \mathscr{A}_{112}(u_1^{(1)})^2 + \mathscr{A}_{222}(u_2^{(1)})^2 + 2\mathscr{A}'_{22}u_2^{(1)}\Lambda^{(1)} + \mathscr{A}^2_2\varepsilon^{2(2)}|^C = 0 \end{aligned}\right\} \tag{E}$$

and the critical state equations (7.31) are

$$\left.\begin{aligned} \mathscr{A}_{112}x_2 u_1^{(1)} + \mathscr{A}_{112}x_1 u_2^{(1)} + \mathscr{A}'_{11}x_1\Lambda^{(1)} + \mathscr{A}'_{11}x_1\sigma^{(1)}|^C = 0 \\ \mathscr{A}_{112}x_1 u_1^{(1)} + \mathscr{A}_{222}x_2 u_2^{(1)} + \mathscr{A}'_{22}x_2\Lambda^{(1)}|^C = 0 \end{aligned}\right\} \tag{C}$$

which take the special form of equations (7.33) for m-fold criticality,

$$\left.\begin{aligned} \mathscr{A}_{112}u_2^{(1)} + \mathscr{A}'_{11}\Lambda^{(1)} + \mathscr{A}'_{11}\sigma^{(1)}|^C = 0 \\ \mathscr{A}_{112}u_1^{(1)}|^C = 0 \\ \mathscr{A}_{222}u_2^{(1)} + \mathscr{A}'_{22}\Lambda^{(1)}|^C = 0 \end{aligned}\right\} \tag{mC}$$

Finally, the extra equation (7.35), specifying secondary bifurcation as opposed to limit points, becomes

$$\mathscr{A}'_{11}x_1 u_1^{(1)} + \mathscr{A}'_{22}x_{22} u_2^{(1)}|^C = 0 \tag{B}$$

Table 7.1 First-order analysis of semi-symmetric bifurcation. The appropriate equations for each situation are as shown, subject to the conditions arising from the exclusion of certain control parameters

	Equations	Conditions	Solution	
Post-buckling of the perfect system	**E**	$\sigma^{(1)} = 0$, $\varepsilon^{1(2)} = \varepsilon^{2(2)} = 0$	Direct	⎫
Imperfection-sensitivity on the symmetric section	**E, C**	$\sigma^{(1)} = \varepsilon^{1(2)} = 0$	Direct	⎬ Complete coincidence $\sigma = 0$
Full imperfection-sensitivity	**E, C**	$\sigma^{(1)} = 0$	Generalized imperfection	
Secondary bifurcations	**E, C, B**	$\sigma^{(1)} = 0$	Direct	⎭
Post-buckling of the perfect system	**E**	$\varepsilon^{1(2)} = \varepsilon^{2(2)} = 0$	Direct	⎫
Bifurcations of the perfect system	**E, C**	$\varepsilon^{1(2)} = \varepsilon^{2(2)} = 0$	Direct	
Imperfection-sensitivity on the symmetric section	**E, C**	$\varepsilon^{1(2)} = 0$	Direct	⎬ Near-coincidence
Locus of m-fold critical point (umbilic point)	**E, mC**		Direct	
Full imperfection-sensitivity	**E, C**		Generalized imperfection, generalized loading parameter	
Secondary bifurcations	**E, C, B**		Generalized loading parameter	⎭

Just which of these equations are to be used, which terms—if any—are absent, and how they are to be solved depend precisely on the problem in hand. For example, if we are to consider the post-buckling of the perfect system, $\varepsilon^1 = \varepsilon^2 = 0$, we would use equations (**E**) with $\varepsilon^{1(2)} = \varepsilon^{2(2)} = 0$, and with $\sigma^{(1)} = 0$ if we are to be concerned just with complete coincidence. The significant alternatives are here presented in Table 7.1.

The generalized imperfection and generalized loading parameter can be used as aids to the solution of the equations when appropriate, and further simplifications which can also be sometimes exploited include specification of the perturbation parameter s and normalization of the local eigenvector x_j. As an example of the former, it is sometimes useful in imperfection-sensitivity studies[92,93] to specify s as the positive square root of the generalized imperfection, written $\varepsilon^{1/2+}$.

Then, writing ε as a Taylor expansion about C we obtain

$$\varepsilon = \varepsilon^{(1)C}s + \tfrac{1}{2}\varepsilon^{(2)C}s^2 + \tfrac{1}{6}\varepsilon^{(3)C}s^3 + \cdots \tag{7.45}$$

and equating coefficients

$$\varepsilon^{(1)C} = \varepsilon^{(3)C} = \varepsilon^{(4)C} = \cdots = 0$$
$$\varepsilon^{(2)C} = 2 \tag{7.46}$$

Sometimes other interpretations of s may be useful, but we must ensure that s is capable of describing all possible solutions; for this reason it may be left undefined and solutions obtained in a rate-space $u_i^{(1)} - \Lambda^{(1)}$, say, although these become the path tangents when mapped directly into the corresponding coordinate space $u_i - \Lambda$.[1] Normalization of the local eigenvector involves simply setting one of the x_i to unity; thus $x_1 = 1$ takes care of all possibilities except $x_1 = 0$, which must be explored as a special case.[93]

Numerical search procedure

This section might be omitted by all readers except those with an interest in the details of the imperfection-sensitivity analysis.

A polar representation of controls may be entirely appropriate for imperfections when we come to plot the final results, but not necessarily for the loads; it is of more interest to view the imperfection-sensitivity, for instance, at a constant value of the splitting parameter σ than a constant angular polar coordinate ϕ. We could translate the second view into the first by fixing the load state and using ε as a dependent variable, but this is contrary to the nature of imperfection-sensitivity plots, which are invariably visualized as imperfection-independent and load-dependent. For a systematic plotting procedure we would clearly like to stick with convention, but since σ is previously specified we are now obliged to introduce a numerical search procedure, necessarily performed on a computer.

Let us thus consider the plotting routine for a typical imperfection-sensitivity surface with the four controls of Λ, σ, ε^1, and ε^2, as shown in Figure 6.11, say. We first fix the values of ε^i, giving θ and ε of the polar representation. Applying transformation (7.37), we can write down the equations for the imperfection ray, and considering for the moment $\sigma = 0$, they can be solved for the set of $\Lambda^{(1)C}$; hence, knowing ε, we obtain a first-order estimate of critical loads.

But when σ is non-zero, the situation is not so simple; the generalized load approach can be used, but we cannot fix the correct load ray without knowing Λ, and it is this that we are seeking in the analysis. We overcome the problem by scanning the complete load space incrementally, using a numerical search involving a reduction in increment size, to home-in on situations where a solution to the equations coincides with the value obtained from the polar transformation. The process, of course, implies greatly increased computer time, and it may therefore be advisable in problems of near-coincidence to force the bifurcations together by introducing a splitting parameter. However, the analyst must be aware of the topological significance of the manoeuvre in the light of the criteria of structural stability and universal unfolding.

8

Buckling of Plates and Shells

We close the book with a brief interpretation of our phenomenological view of elastic buckling, in more practical terms appropriate to real engineering structures. In this we shall entirely ignore plasticity effects, which are clearly important for the kinds of geometries common to offshore structures, for instance. Nevertheless, the phenomenology does highlight a number of quite different, archetypal forms of structural behaviour, each with its own distinctive post-buckling character.

An elastic strut, for example, has very little post-buckling stiffness. This is explained, not by the rigorous large deflection treatment of our earlier work,[1] but by a simple discussion involving first- and second-order energy contributions. The argument, when extended to the unstiffened flat plate restrained on all four edges, accounts for the relatively stiff nature of the post-buckling response. Interactions between two different modes of buckling, leading to a considerable release of membrane energy, go some way to explaining the highly unstable post-buckling and severe imperfection-sensitivity of the cylindrical shell under axial load. Similarly, interactions between local and overall buckling can destabilize the response of a stiffened plate, or further destabilize that of a stiffened shell. In each case of interactive buckling considered, the parabolic umbilic catastrophe is the key underlying phenomenon.

The arguments are based on modal descriptions of deflected shape, as in conventional Rayleigh–Ritz analyses. The elimination of passive coordinates, in the diagonalized format appropriate to such formulations, is found to provide a powerful extra tool for investigating buckling in one (or more) modes under the (passive) influence of (possibly many) other higher modes. Symmetry and optimization are seen to be constantly recurring themes;[46] the former, in several different ways, can sprinkle the potential energy function with zero coefficients,[81] while the latter can draw bifurcations together to induce interactions.

In most cases, the qualitative description of the buckling process is backed up by quantitative evaluation of key energy coefficients. The way is then clear to asymptotically exact results, for all the structural geometries covered, using the techniques of Chapter 7 together with an extension to the parabolic umbilic form.[102] By and large, this is not pursued here.

8.1 ANALYSIS USING PRINCIPAL COORDINATES

In the following treatment of struts, plates, and shells, the contributing bifurcations all arise on a linear fundamental equilibrium path, defining the uniformly compressed state. Of course, some structures exhibit non-linear fundamental paths, but such problems are not to be considered here; we do, however, refer to the arch formulation of Chapter 5, where a highly non-linear fundamental response is modelled by a linear counterpart, with some analytical success.

In the earlier chapters we have, in the main, used the notation $V(q_i, \Lambda^j)$ for the potential function, whatever its properties. The exception is the rigorous bifurcation analysis of Chapter 7 where, following our earlier work,[1] special properties of particular functions are recognized by special symbols, $\mathscr{W}$, $\mathscr{A}$, etc. Adopting this policy here we denote the potential function by $A(u_i, \Lambda^j)$, where u_i represents a set of n incremental *principal* coordinates, measuring mode amplitudes from the uniformly compressed state, and Λ^j is a set of h control parameters; the well-known orthogonal properties of buckling modes render A diagonalized.

The elimination of passive coordinates, in the non-diagonalized format of Chapter 7, allows some freedom in the choice of active and passive coordinates, the only restriction being that the submatrix of quadratic coefficients relating to the passive coordinates is non-singular at the critical point under consideration. Here, however, the m *active* coordinates are specifically those involved in the buckling process and are uniquely defined, the remaining $n - m$ being the *passive* coordinates. Latin subscripts are again exclusively reserved for the actives, the passives being written with Greek subscripts. As before, we make direct use of the passive equilibrium equations

$$\frac{\partial A}{\partial u_\alpha} = 0$$

to introduce a new m-degree-of-freedom potential function defined by the condition

$$\mathscr{A}(u_i, \Lambda^j) \equiv A[u_i, u_\alpha(u_i, \Lambda^j), \Lambda^j], \quad \text{at } \frac{\partial A}{\partial u_\alpha} = 0$$

The new $\mathscr{A}$-function defines the energy level on an *activity hypersurface*, in the full $u_i - u_\alpha - \Lambda^j$ space, carrying a complete description of the system behaviour.[1]

A and $\mathscr{A}$ are both expressed as Taylor series about the fundamental uniformly compressed equilibrium state F, so interest then naturally centres on partial derivatives of the two functions with respect to u_i, u_α, and Λ^j, as appropriate, evaluated at F. As before, a perturbation scheme can write the derivatives of the new function in terms of those of the old. Equilibrium and stability conditions pass over unchanged to the new function,[1] so analysis can be performed entirely in terms of $\mathscr{A}$. We find the relationships between the two alternative formulations of particular interest.

$$\mathscr{A}(u_i, \Lambda^j) = A(u_i, u_\alpha, \Lambda^j)\Big|^{A_\alpha = 0}$$

$$\underset{m}{\nearrow} \quad \underset{h}{\searrow} \quad \underbrace{\qquad}_{n}$$

$$\mathscr{A}^F_{ijk} = A^F_{ijk}, \qquad \mathscr{A}^{jF}_i = A^{jF}_i, \qquad etc.$$

$$\mathscr{A}^F_{ijkl} = A^F_{ijkl} - \sum_{\alpha=m+1}^{n} \frac{1}{A_{\alpha\alpha}} \left(A_{\alpha ij} A_{\alpha kl} + A_{\alpha ik} A_{\alpha jl} + A_{\alpha il} A_{\alpha jk} \right)\Bigg|^F$$

for $\quad m = 2:$

$$\mathscr{A}^F_{1111} = A^F_{1111} - 3 \sum_{\alpha=3}^{n} \frac{1}{A_{\alpha\alpha}} \left(A_{\alpha 11} \right)^2 \Bigg|^F$$

$$\mathscr{A}^F_{1122} = A^F_{1122} - \sum_{\alpha=3}^{n} \frac{1}{A_{\alpha\alpha}} \left[A_{\alpha 11} A_{\alpha 22} + 2 \left(A_{\alpha 12} \right)^2 \right] \Bigg|^F$$

Figure 8.1 The diagonalized elimination of passive coordinates scheme

Figure 8.1 gives a schematic outline of the process, starting with the general definition of $\mathscr{A}$, and then presenting some significant derivatives, including the important fourth derivatives for $m = 2$. Here subscripts denote partial differentiation with respect to the corresponding coordinate u_i or u_α, and lower-case superscripts denote partial differentiation with respect to the corresponding control parameter Λ^j, as before.

This completes our brief guide to the diagonalized elimination of passive coordinates. The most significant qualitative effect appears in the quartic coefficients, $\mathscr{A}^F_{ijkl}$ and A^F_{ijkl}, of the Taylor expansions. Here the direct correspondence apparent for the lower-order coefficients breaks down, and $\mathscr{A}^F_{ijkl}$ exhibits a contamination from the passive contribution, comprising cross-term cubics, $A^F_{\alpha ij}$, etc. and the stability coefficients, $A^F_{\alpha\alpha}$. We now have a significantly different result from an m-degree-of-freedom Rayleigh–Ritz formulation which suppresses the passive coordinates. The contamination invariably has a de-stabilizing effect, as we shall see.

8.2 SYMMETRY AND OPTIMIZATION

We see from the earlier chapters that nominally perfect structural models often exhibit physical symmetries, leading directly to associated symmetries in the underlying potential function.[81] The arch of Chapter 5, reshown here in Figure 8.2, is a simple one-degree-of-freedom example illustrating the general points.

There is a tendency for physical symmetries to be retained on the fundamental path but broken by the post-buckling response; this is the case for symmetry about the vertical centre-line of the arch, for example. But frequently a certain symmetry is exhibit by the post-buckling, namely that equal and opposite values

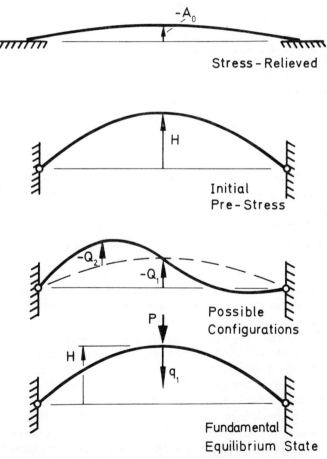

Figure 8.2 Single-degree-of-freedom, inextensional, elastic arch model

of the critical coordinate—Q_2 in the case of the arch—give identical shapes, one being the mirror image of the other reflected about the axis of symmetry, and hence identical energy levels. We refer to this as *symmetry in Q_2*. All odd powers of Q_2 in the Taylor expansion of $A(Q_2, P)$ must vanish, and so

$$A^F_{222} = 0$$

Not surprisingly, the system thus exhibits a symmetric bifurcation.

Of course, symmetry considerations alone cannot predict whether a bifurcation point C is unstable-symmetric, as with the arch, or stable-symmetric. For this we need the fourth derivative A^C_{2222}. For the arch, with the introduction of further degrees of freedom related to Fourier harmonics of higher order than Q_1 and Q_2, the corresponding $\mathscr{A}^C_{2222}$ derivative is liable to contamination from cubics, $A^C_{\alpha 22}$, according to the scheme of Figure 8.1; the appropriate terms in the Taylor expansion cannot be eliminated by symmetry arguments if Q_α represents a waveform that is symmetric about the vertical centre-line. We expect, from the experimental evidence discussed in Chapter 5, that the effect is slight.

It is, of course, the very symmetry of the perfect system that provokes the bifurcational response: asymmetric branching is comparatively rare in elastic buckling. Symmetry-breaking imperfections may now be crucial, as discussed in Chapter 4. Thus a physically symmetrical structure, while perhaps the most natural solution for the designer, may in itself represent an unobtainable optimum in practice.

The general trend of structural optimization also carries a further destabilizing effect, in the call for bifurcations associated with different modes of buckling to approach and coalesce.[1] As an example, the spherical shell—clearly a highly optimized structural form for carrying an external pressure—exhibits a multiplicity of coincident critical loads together with extreme imperfection-sensitivity; the cylinder under axial load responds similarly, again a highly optimized structure. We are thus obliged to take account of possible mode interactions.

In problems of interactive buckling three separate symmetry criteria can be usefully applied.[81] First, is the system symmetric in each mode independently? Second, is it symmetric in one mode in the presence of another? Third, only recently applied to structures,[103,114] is symmetry retained for deflection in all relevant modes simultaneously? These can eliminate some or all cubic terms in the appropriate potential energy expansion, and hence in different combination lead to quite different classifications of the buckling phenomena.

Here we shall concentrate mainly on two interacting modes ($m = 2$), although the contribution of higher modes can be continuously monitored via the elimination of passive coordinates. We show that the bifurcational manifestation of the parabolic umbilic catastrophe, the paraclinal point of bifurcation, is particularly pertinent both, to local and overall interaction in stiffened structures, and to the axially loaded cylinder. Paraclinal branching exhibits symmetry in each of the contributing modes when considered separately, but breaks the symmetry when the modes are combined. Many problems of two-mode interactive buckling seem to satisfy this specific symmetry criterion, a point that has recently been taken up in a mathematical context by Golubitsky et al.,[115] who refer to such problems as exhibiting 'hidden' symmetries.

8.3 THE EULER STRUT

The Euler strut and its elastica certainly represent one of the archetypal forms of structural buckling. It is, however, perhaps unfortunate that it is so often used to introduce stability concepts to undergraduates, since it carries very little post-buckling stiffness. The critical state is thus a close approximation to neutral equilibrium, although the catastrophe theory philosophy tells us that the perfectly flat surface is topologically 'structurally unstable', as discussed in Chapter 3, and therefore cannot arise in the real world; it is, of course, important whether this state is a minimum or otherwise since the system would snap dynamically from the latter but not the former. Nevertheless, the post-buckling effects are considerably less significant for struts than for other structural forms such as plates and cylinders.

The buckled, pin-ended strut was studied over the large deflection range by Euler[6] in 1744, seemingly the first non-linear treatment of a bifurcating system. A perturbation study deep into the post-buckling regime is used for comparison between different formulations in our earlier book,[1] and a single-degree-of-freedom Rayleigh–Ritz model—leading to the exact post-buckling path curvature—is presented here in Chapter 4. In this section we shall draw on these earlier results, but include little in the way of analytical details.

The studies reveal that the membrane energy, associated with stretching of the neutral axis, can be entirely ignored; essentially the same results, notably the famous Euler critical load

$$P^C = \frac{\pi^2 EI}{L^2} \tag{8.1}$$

but also those connected with the post-buckling, can be obtained from an extensional or an inextensional model. The latter, being the simpler, has been employed in all our strut formulations. Thus, for a full analysis, we need merely the strain energy of bending

$$U = \tfrac{1}{2} EI \int_0^L (\ddot{w}^2 + \ddot{w}^2 \dot{w}^2 + \ddot{w}^2 \dot{w}^4 + \cdots) dx \tag{8.2}$$

and the end-shortening

$$\mathscr{E} = \int_0^L (\tfrac{1}{2}\dot{w}^2 + \tfrac{1}{8}\dot{w}^4 + \tfrac{1}{16}\dot{w}^6 + \cdots) dx \tag{8.3}$$

as derived in Chapter 2. We note that these are used again in Chapter 5 to determine the strain energy and the constraint condition of the inextensional arch.

In a typical strut analysis only the first term of each of these expansions will be required by a linear eigenvalue study to reveal the critical load: on the other hand, the post-buckling path curvature needs the second, higher-order term of each. It is clear that, should the curvature somehow involve the first-order variation, it would represent a much more profound influence than this purely second-order effect. This explains the relative insignificance of strut and ring post-buckling,[1] when compared with arches, plates, and cylinders; in each of the latter we can identify a significant first-order component to the curvature term.

For example, the quartic energy term responsible for the post-buckling and imperfection-sensitivity of the arch of Chapter 5 arises directly from the leading strain energy term above, via the constraint condition. Similarly, in the plate analysis which follows, a quartic term of membrane energy is generated from the leading term of the $\mathscr{E}$ expansion above, via the elimination of a passive coordinate u_3.

8.4 POST-BUCKLING OF A COMPRESSED PLATE

We move on to consider the classic buckling problem of a long elastic plate subjected to an end-compression P, as shown in Figure 8.3. When all four edges are simply supported the buckle pattern is known to be sinusoidal in both the transverse and longitudinal directions, with a single half-wave across the width and a number of half-waves along the length.[116] Here, following Koiter and Pignataro,[117] we leave the transverse variation unspecified, so the approach is valid for a wide range of different boundary conditions along the longitudinal edges.

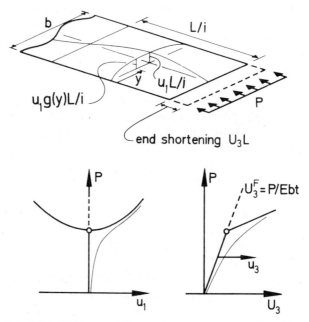

Figure 8.3 The long axially compressed plate, and its typical response

Such problems are usually tackled using the von Karman large-deflection equations.[107,114] We take here an alternative energy formulation in which transverse and shear membrane stresses are ignored, again following Koiter,[117] thereby generating a lower bound to the load level. It is supposed that the results of a linear eigenvalue analysis, for the critical load, are available, and we concentrate here on non-linear effects associated with the membrane energy. Thus, only the quadratic terms of the bending energy are included, as in a linear eigenvalue study, and these are conveniently replaced by the release of membrane energy at the buckling point itself.[116,117] This is entirely consistent with the von Karman approach, which ignores non-linear terms in the strain–displacement relations for the bending stresses but not for the membrane stresses.

The formulation has two degrees of freedom, the amplitude of a single buckling mode u_1 and the end-displacement of the load U_3; it is assumed that the latter is constant across the width, corresponding to compression between rigid platens. We note that U_3 comprises two effects, a pure compression and an end-shortening due to buckling, both of which vary across the width for moderately large out-of-plane deflections. Thus U_3 has a non-trivial fundamental path component, and the analysis illustrates both an incremental-coordinate transformation to allow for this and the elimination of passive coordinates. The extension to n modal degrees of freedom is discussed in the following section.

The out-of-plane deformation is taken to be

$$w = u_1 g(y) \frac{L}{i} \sin \frac{i\pi x}{L} \tag{8.4}$$

which exhibits i half-waves over the length L of the plate and an unspecified variation $g(y)$ across the width. Inextensional deformation, if admissible, would be associated with an end-shortening due to buckling, given by

$$\mathscr{E} = \frac{1}{2} \int_0^L \dot{w}^2 \, dx = \tfrac{1}{4}\pi^2 g^2 L u_1^2 \tag{8.5}$$

to a first order, from the leading term of the expansion (8.3) above. With an actual end-shortening of $U_3 L$ an axial strain is set up in the plate. Taking this to be constant through the length, reasonable in the absence of transverse and shear stresses, it must then be given by

$$\varepsilon_x = (\mathscr{E} - U_3 L)/L = \tfrac{1}{4}\pi^2 g^2 u_1^2 - U_3 \tag{8.6}$$

The membrane strain energy is given simply by $\tfrac{1}{2}E\varepsilon_x^2$, integrated through the volume of the plate.

The plate buckling process, in the small-deflection view of a linear eigenvalue analysis, involves a transfer from membrane strain energy to bending, the critical point defining the transition state at which the system finds it equally easy to store energy in either form. It is thus clear that, as the structure moves into its buckling displacement at the critical load P^c, the bending energy stored exactly equals the membrane energy released, the release being a result of the difference between actual (buckled) length and chord (pre-buckled) length under fixed end conditions;[116,117] this, to a first order, is the same as the end-shortening under inextensional conditions of equation (8.5) above.

The bending energy is thus given by $(P^C/b)\mathscr{E}$, integrated across the width b of the plate. Total potential energy being strain energy less the work done by the load, we thus obtain the potential function

$$V = \tfrac{1}{2}ELt \int_0^b (\tfrac{1}{4}\pi^2 g^2 u_1^2 - U_3)^2 \,\mathrm{d}y + \frac{1}{4}\frac{L}{b}P^C\pi^2 u_1^2 \int_0^b g^2 \,\mathrm{d}y - PU_3 L \qquad (8.7)$$

where t is the plate thickness.

The fundamental path, of course, involves just pure compression, and is given by

$$u_1^F = 0, \quad U_3^F = P/Ebt \qquad (8.8)$$

We define a new incremental coordinate u_3, measuring the additional end-shortening entirely due to buckling, with the uniformly contracted component removed, by the simple transformation

$$U_3 = U_3^F(P) + u_3 = \frac{P}{Ebt} + u_3 \qquad (8.9)$$

Substitution into V, and evaluation at the critical point C, then gives the general form

$$A = \tfrac{1}{24}A_{1111}^C u_1^4 + \tfrac{1}{2}A_{113}^C u_1^2 u_3 + \tfrac{1}{2}A_{33}^C u_3^2 + \tfrac{1}{2}(P - P^C)A_{11}^{\prime C} u_1^2 \qquad (8.10)$$

the prime, as before, denoting partial differentiation with respect to the loading parameter P, where

$$A_{1111}^C = \tfrac{3}{4}\pi^4 ELt \int_0^b g^4 \,\mathrm{d}y, \quad A_{113}^C = -\tfrac{1}{2}\pi^2 ELt \int_0^b g^2 \,\mathrm{d}y,$$

$$A_{33}^C = ELtb, \quad A_{11}^{\prime C} = -\tfrac{1}{2}\frac{L}{b}\pi^2 \int_0^b g^2 \,\mathrm{d}y \qquad (8.11)$$

Here constant terms have been ignored, since they inevitably vanish on differentiation with respect to the u_i for equilibrium.

The symmetry arguments, which we apply later to good effect for more complex systems, are here not strictly necessary, the potential function having been obtained explicitly. However, they could have been used to predict the diagonalization ($A_{13}^F = A_{13}^{\prime F} = 0$), along with $A_{111}^F = A_{133}^F = 0$, the potential function clearly being symmetric in u_1 in the presence of u_3. The diagonalization allows u_3 to behave as a passive principal coordinate, although we note that the structure can never exhibit criticality in that mode since $A_{33}^F > 0$ for all values of P. Also, in typical fashion,[1] the incremental transformation has removed P from acting on a linear term in U_3, and obliged it to act on a quadratic term in the buckling mode u_1.

Eliminating u_3 as a passive coordinate we obtain first u_3 as a function of u_1,

$$u_3(u_1) = \frac{1}{2} \frac{d^2 u_3}{du_1^2}\bigg|^C u_1^2 = -\frac{1}{2} \frac{A_{113}}{A_{33}}\bigg|^C u_1^2 = \frac{1}{4} \frac{\pi^2}{b} u_1^2 \int_0^b g^2 dy \qquad (8.12)$$

to a first order, and second, according to the general scheme of Figure 8.1,

$$\mathscr{A}_{1111}^C = A_{1111} - 3\frac{(A_{113})^2}{A_{33}}\bigg|^C = \tfrac{3}{4}\pi^4 ELt\left[\int_0^b g^4 dy - \frac{1}{b}\left(\int_0^b g^2 dy\right)^2\right],$$

$$\mathscr{A}_{11}^{'C} = A_{11}^{'C} = -\tfrac{1}{2}\frac{L}{b}\pi^2 \int_0^b g^2 dy \qquad (8.13)$$

These can be substituted into the general form for the curvature of the post-buckling path at a symmetric point of bifurcation, given by equation (4.30) of Chapter 4,

$$P^{(2)C} = -\frac{\mathscr{A}_{1111}}{3\mathscr{A}_{11}'}\bigg|^C = \tfrac{1}{2}\pi^2 Ebt\left[\frac{\int_0^b g^4 dy}{\int_0^b g^2 dy} - \frac{1}{b}\int_0^b g^2 dy\right] \qquad (8.14)$$

to construct a first-order approximation to the path

$$P = P^C + \tfrac{1}{2}P^{(2)C}u_1^2 \qquad (8.15)$$

valid for 'moderately large' deflections.

Of course, the model is crude, but it is simple, and it does incorporate the major feature of plate post-buckling—a membrane stretching as the plate is obliged to take on doubly-curved shapes with ever-increasing Gaussian curvature.[118] As an example, with simply supported edges, the model exhibits an effective stiffness in the post-buckling range of one third of the pre-buckled value; Koiter points out that with no in-plane restraint along the longitudinal edges this represents a close lower bound to the correct value of 0.408.[117] The neglecting of transverse and shear membrane stresses can be seen as replacing the plate with an infinite number, across the width, of infinitely thin strips which span the length. Thus, as far as the membrane (but not the bending) energy is concerned, there is no transfer of stress between strips, although they are constrained to act together during buckling.

The influence of higher modes

The above analysis starts with just two degrees of freedom, and reduces to one by eliminating the end-shortening u_3 as a passive effect. We now show that the results (8.14) and (8.15) hold for an n-degree-of-freedom formulation, where the extra $n - 1$ degrees correspond to the amplitudes of higher buckling modes; we assume here that the u_3 elimination has been completed, so the end-shortening contamination of all modes is fully taken into account. The argument hinges on the third symmetry criterion discussed earlier, the potential function clearly being symmetric when all modes are reversed simultaneously. This is used to eliminate all relevant cubics, leaving the contamination of the buckling mode by higher modes of a rather mild form.

We suppose that the perfect system is represented by the potential function $A(u_i, \Lambda)$, where Λ is a loading parameter. The third symmetry condition is then expressed by

$$A(u_i, \Lambda) = A(-u_i, \Lambda) \tag{8.16}$$

We note that, however we chose the active coordinates, the same condition must extend to the $\mathscr{A}$ function, so

$$\mathscr{A}(u_i, \Lambda) = \mathscr{A}(-u_i, \Lambda) \tag{8.17}$$

after reversing all (active and passive) modes.

Starting with $m = 1$ and taking each mode in turn, equation (8.17) leads to

$$\mathscr{A}^C_{iii} = 0 \tag{8.18}$$

as before, after considering a Taylor expansion of $\mathscr{A}$. Thus we can say immediately,

$$A^C_{iii} = 0 \tag{8.19}$$

noting that we can interchange freely between the $\mathscr{A}$ and A representations, since for a diagonalized formulation cubics are uncontaminated by passive effects according to the scheme of Figure 8.1.

We now set $m = 2$, and take the modes in pairs, eliminating the remaining $n - 2$ as a passive effect. Denoting the two active modes as u_i and u_j, reversing them both and equating the cubic contributions, and remembering that $\mathscr{A}^C_{iii} = 0$, we have

$$\mathscr{A}^C_{iij} u_i^2 u_j + \mathscr{A}^C_{ijj} u_i u_j^2 = 0 \tag{8.20}$$

This must hold for any ratio u_i/u_j, and so

$$\mathscr{A}^C_{iij} = A^C_{iij} = 0 \tag{8.21}$$

By similar arguments with three modes we obtain

$$\mathscr{A}^C_{ijk} u_i u_j u_k = 0 \tag{8.22}$$

which again must hold however we emerge from C in $u_i - u_j - u_k$ space. Thus in general,

$$\mathscr{A}^C_{ijk} = A^C_{ijk} = 0 \tag{8.23}$$

and all cubics are seen to vanish.

We thus destroy the possibility of any contamination of an active coefficient $\mathscr{A}^C_{ijkl}$, according to the scheme of Figure 8.1, whatever the choice of passive coordinates. The derivatives (8.13), and the curvature expression (8.15), thereby remain unaffected by the inclusion of higher modes of buckling; this continues to apply for n very large, even infinite. However, the range of validity of the curvature could be much reduced by higher bifurcations close to C on F; this enters the realm of interactive buckling, to be discussed later. The conclusions can be applied equally well to struts and plates, and similar reasoning might also be used to good effect for other structural forms such as cylinders.

We note finally that we have been able to come to these conclusions without reference to the second symmetry criterion—an altogether stronger statement referring to symmetry in one mode in the presence of another. This has frequently been used in the literature to define *double-symmetric* systems,[1,91] leading to bifurcational manifestations of the *double-cusp catastrophe*. The second criterion is not satisfied in general for plates or struts, breaking down if both modes carry an odd number of waves longitudinally.

8.5 INTERACTIVE BUCKLING OF STIFFENED STRUCTURES

The analysis of plate post-buckling in the previous section indicates just how the presence of a passive mode can adversely affect the post-buckling—by contamination of the fourth derivative $\mathscr{A}^C_{1111}$. The phenomenon has one essential degree of freedom, and u_3 can be eliminated as a passive effect, precisely because the stability coefficient $\mathscr{A}^F_{33}$ is nowhere zero. For some stiffened structures, an overall mode of buckling can contaminate a local mode in a similar way, with the important difference that its stability coefficient does vanish, or is at least very small, at the local buckling load.[103,117] The effects of the contamination is then most profound, drastically reducing the appropriate fourth derivative, and no single-degree-of-freedom treatment suffices. It is this phenomenon, marked by two (or more) critical points of bifurcation occurring at, or near, the same load, that we refer to as interactive buckling.[119-25]

Let us therefore consider the stiffened plate under compressive load Λ shown at the top of Figure 8.4. Here we have two forms of buckling, the overall Euler mode of the middle diagram and the local plate mode, similar to that of the previous section, at the bottom. The Euler mode exhibits no transverse y-variation and is thus effectively a column failure, although it also approximates the response of a wide plate.[98] We follow the analysis of reference 103, which itself follows Koiter's continuum approach,[117] and describe the modes by

$$w(x, y) = u_1 g(y) \frac{L}{i} \sin \frac{i\pi x}{L}, \quad W(x) = u_2 L \sin \frac{\pi x}{L} \tag{8.24}$$

i being the number of half-waves of the local mode along the length L of the plate. With two essential degrees of freedom we are, of course, entering the province of the umbilic catastrophes of Chapter 7.

Here we can make considerable use of the symmetry properties of the system. We start by asking the question posed by the first symmetry criterion—is the potential function symmetric in each mode independently? This is clearly the case for the local mode, as in the previous section. It is also approximately true for the

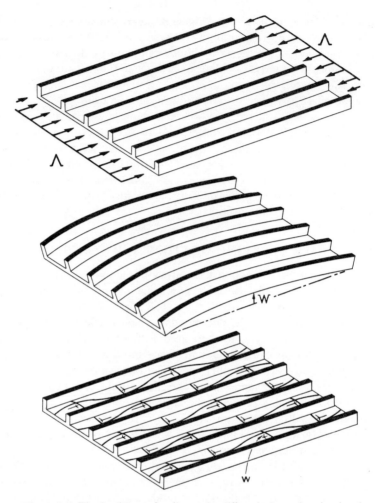

Figure 8.4 The buckling of a discretely stiffened plate, showing local and overall modes

overall mode, for without the interactive effects the discretely stiffened plate can successfully be replaced with a perfectly orthotropic counterpart, smearing out the effect of the stiffeners.[103] We thus have

$$\mathscr{A}^C_{111} = \mathscr{A}^C_{222} = 0 \tag{8.25}$$

for a two-fold $(m = 2)$ potential function $\mathscr{A}(u_1, u_2, \Lambda)$ of the perfect system. We see again that we are entitled to interchange freely between one- and two-dimensional forms for these derivatives, since by the scheme of Figure 8.1 cubics are uncontaminated by passive effects.

We next pose the second symmetry question—is the potential function symmetric in one mode in the presence of the other? With i an even-number, symmetry is guaranteed in u_1, even in the presence of u_2, by mirror-image reflection about the centre-line (like the arch). This gives a diagonalized potential function, and

$$\mathscr{A}^C_{122} = 0$$

But the opposite is clearly not the case. In the presence of the local mode u_1, reversing u_2 must lead to a quite different energy level. It is crucial to the local buckling response, whether the overall mode acts so as to enhance, or detract from, the compressive effect, with deformation away from, or towards, the stiffeners. This asymmetry leads to a marked imperfection-sensitivity, and is directly related to the cross-term cubic $\mathscr{A}^C_{112}$. We can determine this for a highly simplified model of the interaction as follows.

Following Koiter and Pignataro,[117] as earlier, we start by neglecting transverse and shear membrane stresses. We next assume that the neutral surface is inextensional, which is allowable because the cross-term cubic of interest remains uncontaminated by passive, end-shortening effects. A typical section through a panel between stiffeners is shown in Figure 8.5.

We take first just local mode deformation, as shown at the top of the figure. Inextensibility of the neutral surface, and zero curvature in the overall mode, means that the panel must undergo longitudinal stretching. To a first order this change in length is

$$\delta L = \tfrac{1}{2} \int_0^L \dot{w}^2 dx \tag{8.26}$$

being the same as the end-shortening under inextensional conditions, as discussed earlier in the plate analysis. Assuming that this sets up a constant axial strain through the length, we have $\varepsilon_x = \delta L / L$ in the deformed panel.

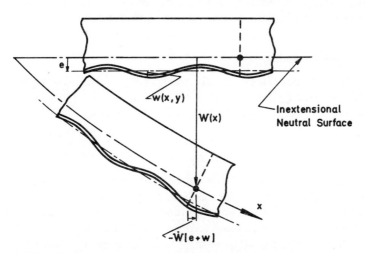

Figure 8.5 Simultaneous buckling in both local and overall modes of the stiffened plate, at a typical section $y = $ constant

The stretching must involve some axial displacement, but we now assume that after this has taken place, during the overall deformation, plane sections remain plane and shear lag is ignored. Approximating for small angles, we thus have a further displacement in the x-direction of $-\dot{W}(e+w)$, e being the panel eccentricity, as shown. This sets up the additional axial strain

$$\varepsilon_x = -\partial \dot{W}(e+w)\partial x = -\dot{W}'(e+w) - \dot{W}\dot{w} \qquad (8.27)$$

Substituting the assumed forms and performing the differentiations, we obtain the total axial strain in the panel as

$$\varepsilon_x = \tfrac{1}{4}\pi^2 g^2 u_1^2 + \frac{e}{L}\pi^2 u_2 \sin\frac{\pi x}{L} + \pi^2 g u_1 u_2 \left(\frac{1}{i}\sin\frac{\pi x}{L}\sin\frac{i\pi x}{L} - \cos\frac{\pi x}{L}\cos\frac{i\pi x}{L}\right) \qquad (8.28)$$

Having ignored transverse and shear stresses, we obtain the membrane energy of a single panel by integrating $\tfrac{1}{2}E\varepsilon_x^2$ over its volume. The cross-product of the first two terms gives

$$\mathscr{A}_{112}^C = \pi^3 E t e \int_0^b g^2 \mathrm{d}y \qquad (8.29)$$

the cubic of interest for the single panel of breadth b and thickness t.

It should be noted that the deflection pattern of Figure 8.5 is only strictly valid for vanishingly small displacements; the local mode w inevitably involves an axial component, and this conflicts with plane sections remaining plane in the overall mode W. The situation is unchanged by the introduction of neutral surface extensibility as a passive effect,[103] as in the earlier plate analysis.

The third symmetry question—is the potential function symmetric in both modes simultaneously? (answer, no)—of course provides no extra information here. However, it can be used, as in the earlier plate example, to eliminate all cubics associated with interactions between different local modes.[103]

Parabolic umbilic catastrophe

The vanishing of all cubics except the single cross-term $\mathscr{A}_{112}^C$ indicates that the two-fold compound bifurcation C is a parabolic umbilic point. The potential function for the perfect system can thus be written

$$\mathscr{A} = \tfrac{1}{2}\mathscr{A}_{112}^C u_1^2 u_2 + \tfrac{1}{24}\mathscr{A}_{2222}^C u_2^4 + \tfrac{1}{4}\mathscr{A}_{1122}^C u_1^2 u_2^2 + \tfrac{1}{24}\mathscr{A}_{1111}^C u_1^4$$
$$+ \tfrac{1}{2}(\Lambda - \Lambda^C)(\mathscr{A}_{11}'^C u_1^2 + \mathscr{A}_{22}'^C u_2^2) + \tfrac{1}{2}\sigma\mathscr{A}_{11}'^C u_1^2 + \text{higher-order terms} \qquad (8.30)$$

where σ is a control parameter introduced to examine the separation of the two contributing bifurcations on the fundamental path, as discussed in Chapter 7.

An asymptotic study of this potential function along the lines of the bifurcation analysis of Chapter 7 reveals some surprises.[101,102] The mathematical concept of determinacy suggests that the quartic $\mathscr{A}_{2222}^C$ term should join the cubic $\mathscr{A}_{112}^C$ as a first-order effect, while the remaining quartics $\mathscr{A}_{1122}^C$ and $\mathscr{A}_{1111}^C$ are relegated to the higher-order terms, giving rise to the plots of Figure 6.21. This idea will be unfamiliar to most workers in structural mechanics, and implies that u_1 and u_2 (and ε^1 and ε^2 in an imperfection-sensitivity study) play slightly different asymptotic roles.

It is also worth noting, seeing that $\mathscr{A}^C_{2222}$ is of first-order significance, that it remains uncontaminated by higher local modes of buckling, according to the same symmetry criterion which eliminates the cubic $\mathscr{A}^C_{122}$. The same does not apply to the higher-order $\mathscr{A}^C_{1122}$ term, which may well exhibit, according to the scheme of Figure 8.1, a collection of non-zero $\mathscr{A}^C_{\alpha12}$ contributions associated with a cluster of local modes with critical loads only slightly higher than Λ^C. Similar conclusions might also be drawn for the two-fold cylinder buckling study briefly discussed later.

Finally, to obtain all the derivatives of equation (8.30), neutral surface extensibility must be included. This gives a non-trivial fundamental path component and a passive effect u_3, as in the earlier plate formulation, the latter possibly affecting the entire quartic contribution but not the cross-term cubic $\mathscr{A}^C_{112}$ primarily responsible for the interaction. For the problem of Figure 8.4, in which the overall mode is a column failure, $\mathscr{A}^C_{2222}$ would be positive but small, as discussed earlier in relation to struts, and contamination from u_3 would be negligible by analogy with equations (8.13). The canonical (first-order) form of equation (8.30) can thus be explored with an inextensional neutral surface,[103] although the relative size of $\mathscr{A}^C_{2222}$ suggests that this parabolic umbilic lies in the substrata of, but uncomfortably close to, some higher-order catastrophe. The conclusion cannot be extended to other stiffened structures, for which neutral surface extensibility remains highly significant.

Butterfly catastrophe

A second relatively high-order catastrophe is intimately connected with the potential function of equation (8.30). Let us suppose that local buckling arises first on the fundamental path under increasing load, and that with no significant interaction this is a stable critical state. The interaction develops as σ approaches zero, and the local buckling is somewhere obliged to change from a stable to an unstable form of bifurcation, marking the 'stability limit of the pre-buckled state' discussed by Koiter.[126] This is a butterfly catastrophe point.

It is a relatively simple matter to demonstrate the phenomenon using the elimination of passive coordinates scheme of Figure 8.1. This is done in Section 4.9, which also includes schematic diagrams of the associated equilibrium paths in Figure 4.17; we note here that, although buckling is initiated in the mode u_1, the contamination implies that it quickly develops into the combined mode, an effect that it not shown in the figure.

Stability at the butterfly point is controlled by the sixth-order term $\mathscr{A}^C_{111111}$, the description of which is beyond the scope of this analysis. However, the elimination of passive coordinates may allow such a term to develop from the lower-order contributions of equation (8.30); this interesting point is not pursued here. We note finally that the two phenomena of the parabolic umbilic and butterfly catastrophes appear juxtaposed in this way because of their relative positions in the substrata of the double cusp.

8.6 THE AXIALLY COMPRESSED CYLINDRICAL SHELL

Thin shells, being extremely efficient carriers of load, also exhibit the most severe forms of instability.[1] Of all shell problems, perhaps the most important from a practical point of view, and certainly one of the most complex phenomenologically, is the axially loaded cylinder; this is rivalled in analytical complexity, if not in practical significance, by the externally pressurized sphere. We concentrate here on these two classic problems, identifying for each a typical form with a linear fundamental path, which loses its stability at a high order, multiple, compound bifurcation.

We do not pretend to deal with these two problems in anything more than a perfunctory manner. The careful bifurcational approach which is the theme of this book, taking into account the topological concepts of determinancy and unfolding, is clearly in both cases quite out of the question, the phenomena concerned being of such high order. The closest approach for each is an extensive treatment by Koiter,[12,127,128] whose powerful contributions we briefly discuss. We can, however, in each case, identify a significant two-fold component to the interaction. The symmetry arguments, and the concept of elimination of passive coordinates, suggest that certain limited conclusions could be drawn for the multi-compound problems, but these are not pursued here.

Following Koiter,[12] we consider here only a pure form of cylinder buckling, in which the boundary conditions at the ends of the shell are ignored. The treatment thus only applies to long cylinders, which we assume are restrained against buckling as a column. The shorter the cylinder, of course, the more significant the end-conditions are likely to be, frequently introducing an axisymmetric barrelling into the pre-buckling deformation, for example, leading to a non-linear fundamental path. A few effects of different assumed end conditions are neatly summarized by Hutchinson and Koiter.[47]

The classical solution from shallow shell theory reveals critical modes,[12,129,130] with axial wavelength $2l$ and circumferential wavelength $2b$, related by a half-circle as shown at the top of Figure 8.6. Each mode describes a chequerboard pattern over the complete shell, with sinusoidal variations both axially and circumferentially. Here l_0 is given by

$$l_0 = \pi \sqrt{(Rt/2c)} \tag{8.31}$$

for a cylinder of radius R and thickness t, where c is given by

$$c = \sqrt{[3(1 - v^2)]} \tag{8.32}$$

v being Poisson's ratio. Of course, the infinite smear of wavelengths implied by the half-circle reduces to a discrete number of contributing modes, since each must have an integer number of waves about the circumference; a similar periodicity requirement will also apply axially for shells of finite length.

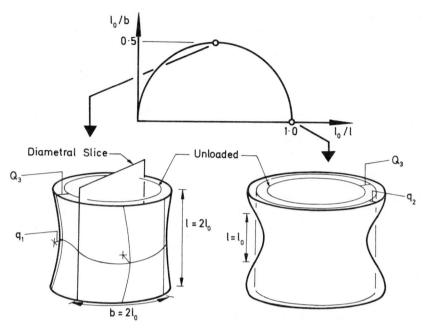

Figure 8.6 The half-circle of contributing modes for the axially loaded cylinder and the two under consideration here

We treat here a strong interaction between two of the contributing modes, shown at the bottom of Figure 8.6. The left-hand pattern, arising at the crown of the half-circle, is a double sine variation with $l = b$; here we show only two full waves circumferentially and one half axially, although in reality a thin shell would buckle in many more circumferential waves, and the pattern is repeated *ad infinitum* along the length. The right-hand mode, from the far right of the half-circle as shown, is the only member of the set not exhibiting a chequerboard pattern; it is axially symmetric with a wavelength of one half of the left-hand mode. We note also that for strong coupling the second mode must be, in a sense, out of phase with the first, its lines of maximum outward deflection rather than its nodal lines coinciding with the circumferential nodal lines of the chequerboard.

Amplitudes of the two modes are denoted by q_1 and q_2 for the chequerboard and axisymmetric patterns, respectively, and we also include a third degree of freedom Q_3, representing a uniform dilation of the cylinder as shown. This latter is clearly associated with the total end-shortening via Poisson's ratio, and thus has very similar properties to the end-shortenings of the plate formulations; it comprises a fundamental component Q_3^F and a further incremental component q_3 entirely due to buckling.

We proceed now to examine the modes by our symmetry criteria. First, we see that the potential function is symmetric in each mode independently; for q_1 we have mirror-image reflection about the vertical diametral slice as shown, while for q_2, repeatability along the length implies symmetry. These statements necessarily

182

hold in the presence of Q_3^F, as we are concerned with symmetries about the fundamental equilibrium state. We thus have

$$\mathscr{A}_{111}^C = \mathscr{A}_{222}^C = 0 \tag{8.33}$$

in the Taylor expansion of the potential function $\mathscr{A}(q_1, q_2, \Lambda)$ about the two-fold compound critical state C, where load $\Lambda = \Lambda^C$.

Next, the potential function is clearly symmetric in q_1 in the presence of q_2. This holds by mirror-image reflection about the diametral slice as before, and, as with Q_3^F, is a direct consequence of the axial symmetry of q_2. Thus

$$\mathscr{A}_{122}^C = 0 \tag{8.34}$$

But we have no similar reason to suppose symmetry in q_2 in the presence of q_1. With q_2 acting in the direction shown in the figure, the inward deflection at the mid-length of the illustrated section, for instance, counteracts the inevitable circumferential stretching of the chequerboard pattern: reverse q_2 and both modes induce circumferential stretching. We would thus expect

$$\mathscr{A}_{112}^C \neq 0 \tag{8.35}$$

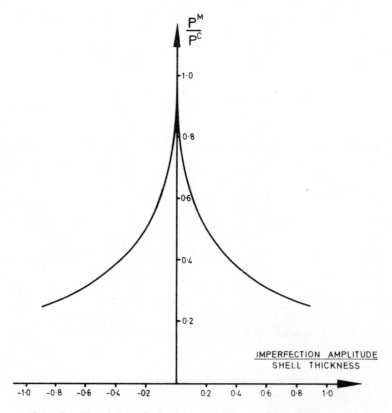

Figure 8.7 The full imperfection-sensitivity of the cylindrical shell, determined by Koiter[12]

So we have the same form of potential function as equation (8.30) for the stiffened structure interaction, although very different values of coefficients might be expected, and C is a parabolic umbilic catastrophe point. The form has been confirmed in a seven-degree-of-freedom formulation by K.A.J. Williams. This involves the removal of axial components of deflection for both modes, and a circumferential component for q_1, using the non-diagonalized elimination of passive coordinates of Chapter 7. The uniform dilation Q_3 is then removed, first by an incremental transformation to get rid of Q_3^F, and second, by eliminating q_3 as a passive coordinate using the diagonalized scheme of Figure 8.1.

Again it must be stressed that this two-fold interaction provides a far from complete picture of cylinder buckling, although it is the basis of Donnell's early, highly stimulating argument for the extreme instabilities observed in experi-

Plate 8.1 A Yoshimura pattern folded from a flat sheet of card. Note that the pattern is of fixed constant amplitude; any attempt to change the amplitude of deformation would result in membrane stretching

ments;[131] it has also been revived by Croll,[132] in a search for lower bounds to the critical loads of imperfect cylinders. The imperfection-sensitivity is left unexplored, but we would not expect the severe form shown in Figure 8.7, drawn from Koiter's analysis of his thesis.[12] This is the outcome of a full m-fold treatment, which highlights the extreme instability of the critical state by the appearance of a number of cross-term cubics apart from $\mathscr{A}^C_{112}$. Each involves an axisymmetric component, however, and interactions between the same chequerboard mode circumferentially out of phase with itself certainly do not appear on a cubic level; these have been used by Poston[5] to argue that m is infinite, determinacy demands terms of infinite order, and unfolding demands an infinite number of control parameters, in the classic cylinder buckling problem. We note finally that the cubics, arising from asymmetries like that of condition (8.35), seem to provide a fundamental mechanism for the breaking of rotational symmetry, apparent also in the spherical shell.[128]

Koiter's second approach of 1963 is to take an imperfection of finite size in the axisymmetric mode, thereby generating a non-linear fundamental path.[127] A secondary bifurcation into a non-symmetric mode can then be treated as a distinct eigenvalue problem, the imperfection having separated the critical points; clearly here the same symmetry-breaking mechanism can play an important part. The analysis reduces the dependence on vanishingly small imperfections of the earlier analysis, although the two are seen to give close agreement, particularly for small imperfections.

We close with a few comments on the large-deflection response. The chequerboard modes of Figure 8.6 are only strictly valid for very small displacements; as they grow, the interactions become so profound that the buckle pattern changes completely to the well-known diamond form. This bears a remarkable similarity to the 'Yoshimura pattern',[133] formed by folding a flat sheet of card as shown in Plate 8.1. Here we have a deformed state with an unstretched neutral surface, although to attain that state the shell must pass through extensional configurations—clearly an unstable process. We note finally that an infinite range of Yoshimura patterns are allowable on a cylinder.

8.7 THE EXTERNALLY PRESSURIZED SPHERICAL SHELL

The thin spherical shell subjected to a uniform external pressure provides a second classic buckling problem featuring a large number of interacting modes. This has much in common with the axially loaded cylinder; in particular, it exhibits both rotationally symmetric and non-symmetric modes, which apparently interact in a similar dramatic way.[128] We concentrate here on rotationally symmetric modes only, drawing attention to a second form of interaction between two of them, giving a rather strange manifestation of the hyperbolic umbilic catastrophe. A similar phenomenon might also arise for cylinders of finite length, but is unlikely to carry quite the significance of the sphere problem; in the latter, the rotationally symmetric modes are unstable in their own right, and the interaction provides a fair approximation of the large-

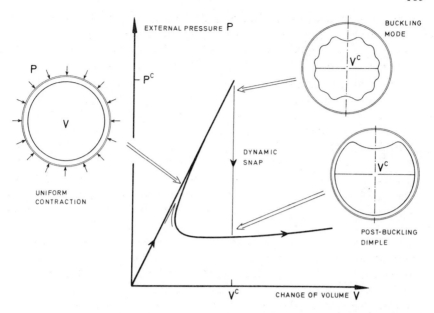

Figure 8.8 The large-deflection response of the spherical shell, showing the constant volume dynamic snap

deflection pattern of Figure 8.8, although in reality the shell may pass through non-symmetric configurations to attain this state.

The early pioneering work of von Karman and Tsien focused attention on the extreme instability of the buckling process for spherical shells,[134] demonstrating that large-deflection configurations exist at pressures considerably below the classical critical value. The complete situation is shown in Figure 8.8. The buckling modes are described by Legendre functions, while the large deflection buckle pattern is usually a single dimple as shown. The figure shows the extreme instability of the sphere; severe dynamic snaps occur even under constant volume control, representing rigid as opposed to dead loading, briefly described in the following section.

The corresponding pressure–volume characteristic for a real shell of radius to thickness ratio $R/T = 19.15$ is shown in Figure 8.9. Here Thompson's theoretical and experimental post-buckling solutions are seen to be in quite good agreement. The 'lower buckling load', representing the minimum pressure in the advanced post-buckling equilibrium solution, is seen to be about one-fifth of the eigenvalue buckling load. More details of this study of sphere buckling are contained in the following section on the laboratory dead and rigid loading of elastic structures.

In 1964 Thompson analysed the rotationally symmetric response of the shell, demonstrating theoretically the instability of the critical state with the appearance of cubic energy terms.[135] This was followed in 1967 by a neat contribution from Hutchinson,[136] who restricted attention to a shallow segment of the shell, ignoring continuity conditions on the back surface, and thereby

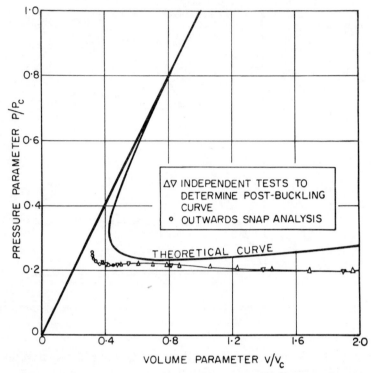

Figure 8.9 Thompson's theoretical and experimental post-buckling curves in the response of a uniformly compressed complete spherical shell

obtained a simple, approximate, but successful, solution. Both analyses were summarized and extended by Koiter in 1969,[128] which remains the only really complete study of the problem, including both symmetric and non-symmetric behaviour. Among Koiter's major findings is that a perturbation approach based on expansion about the compound critical state has little relevant range of convergence; higher-order terms appear with increasing orders of magnitude and therefore cannot be ignored. This seems to imply a severely distorted energy surface.

The two-fold analysis of our earlier work[1] follows, with slight modifications, the rotationally symmetric treatment of Koiter.[128] The range of convergence is, of course, suspect, as discussed above. It does, however, exhibit an unusual property, the 'spherical shell condition', briefly discussed in Chapter 6 (Figure 6.18), and is thus of some interest.

The analysis starts with the removal of the non-trivial but linear fundamental path, representing a pure membrane compression. An expansion in an infinite number of Legendre polynomials then reveals the critical loads

$$\Lambda^n = \frac{1 - v^2}{n(n + 1)} + \alpha n(n + 1) \tag{8.36}$$

for the polynomial of degree n; for a thin shell, n will be large, and we thus consistently neglect n^2 in comparison with unity. The dimensionless loading parameter Λ is here given by

$$\Lambda = \frac{pR(1 - v^2)}{2ET} \tag{8.37}$$

and α is the geometric constant

$$\alpha = \frac{1}{12}\left(\frac{T}{R}\right)^2 \tag{8.38}$$

for a sphere of radius R, thickness T, and external pressure p.

Treating Λ^n as a continuous function of n, we obtain the schematic form of Figure 8.10. We see that the possibility arises of two critical loads, Λ^{t-1} and Λ^t, say, related to Legendre polynomials of degree $t - 1$ and t, coinciding. Moreover, one of the modes is of even order and symmetric about the equator (as at the top of Figure 8.8), while the other is of odd order and skew-symmetric. Here t is assumed to be even.

Careful consideration shows that, somewhat like the arch, the system is symmetric in the mode of odd order (amplitude a_{t-1}), in the presence of the even mode (amplitude a_t). The potential function thus takes the form of the hyperbolic umbilic catastrophe

$$\mathscr{A} = \tfrac{1}{6}\mathscr{A}_{ttt}^C a_t^3 + \tfrac{1}{2}\mathscr{A}_{t-1t-1t}^C a_{t-1}^2 a_t + \tfrac{1}{2}(\Lambda - \Lambda^C)(\mathscr{A}_{t-1t-1}'^C a_{t-1}^2 + \mathscr{A}_{tt}'^C a_t^2)$$
$$+ \text{ higher-order terms} \tag{8.39}$$

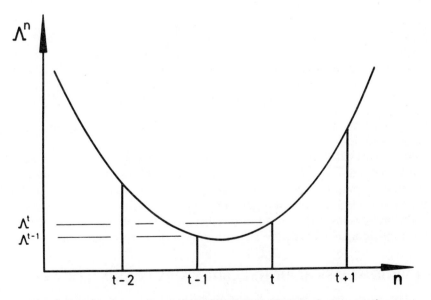

Figure 8.10 The form of equation (8.36), with Λ^n treated as a continuous function of n

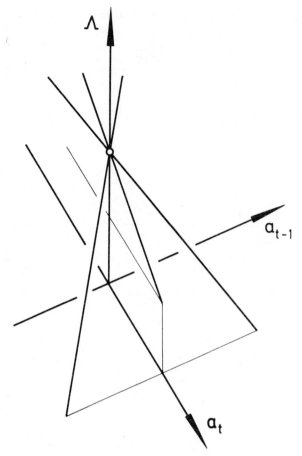

Figure 8.11 The slopes of the post-buckling paths at the double bifurcation for the rotationally symmetric buckling of a complete spherical shell

the analysis[1] giving expressions for the above coefficients which agree with the results of Koiter.[128]

But the particular ratio between the slopes of the coupled and uncoupled paths, shown here in Figure 8.11, suggests that the normal way of plotting the imperfection-sensitivity would improperly unfold the singularity to give the structurally unstable form of Figure 6.18. We refer to this as the *spherical shell condition*[13] but investigate it no further here, merely noting that a structurally stable unfolding would require a fourth dimension of control.

8.8 RIGID AND SEMI-RIGID LABORATORY LOADING DEVICES

We close with a few comments concerning structures and their loading devices, drawing attention to the important difference between *dead* and *rigid* loading.

189

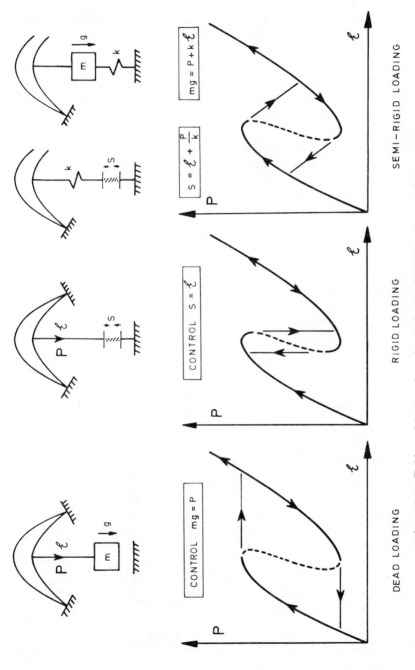

Figure 8.12 Folds and hysteresis cycles under dead, rigid, and semi-rigid loading

While it is usually easy to ensure dead loading of an experimental model, it is, of course, impossible without resort to a complicated control system to impose a strict value of the corresponding deflection $\mathscr{E}$, because of the inherent elasticity of any loading frame. We are then obliged to consider the four situations of Figure 8.12.

In the first arrangement we have dead loading by the imposition of $P = mg$ as the single control parameter, with the energy function $V = U(Q_i) - P\mathscr{E}(Q_i)$. Fold catastrophes arise at extreme values of P, and dynamic snaps occur at constant P as indicated. The area enclosed by the path and the snaps represents the energy lost in the hysteresis cycle.

In the second arrangement we have *idealized* rigid loading by the imposition of $\mathscr{E} = S$, where S is the contraction of a perfectly rigid screw device. The corresponding deflection $\mathscr{E}$ is now our control parameter; folds arise at extreme values of $\mathscr{E}$ and dynamic snaps are at constant $\mathscr{E}$, as indicated. Again the enclosed

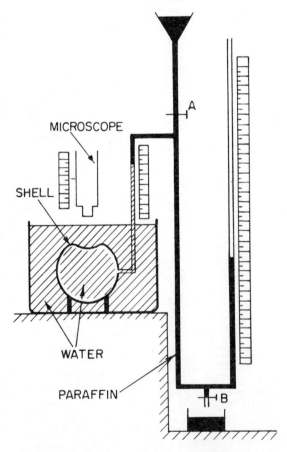

Figure 8.13 An experimental loading device for the volume control of a complete spherical shell[138]

area denotes the energy lost, and we see that it is considerably smaller than the first hysteresis cycle.

The third arrangement shows a practical approach to rigid loading with a spring of stiffness k representing the elasticity of the whole testing device. This spring, plus the perfectly rigid screw, represents a valid model of the majority of *semi-rigid* loading devices. Our control parameter is $S = \mathscr{E} + P/k$, so that we have

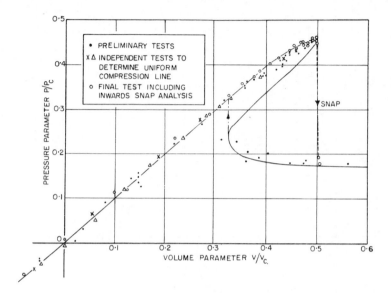

Figure 8.14 Experimental pressure–volume curves showing the results of an inwards-snap analysis[138]

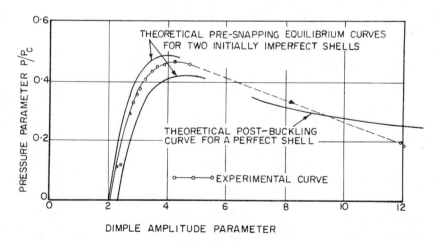

Figure 8.15 Experimental pressure–amplitude curves showing the results of an inwards-snap analysis: comparison is made to a theoretical computer investigation[138]

192

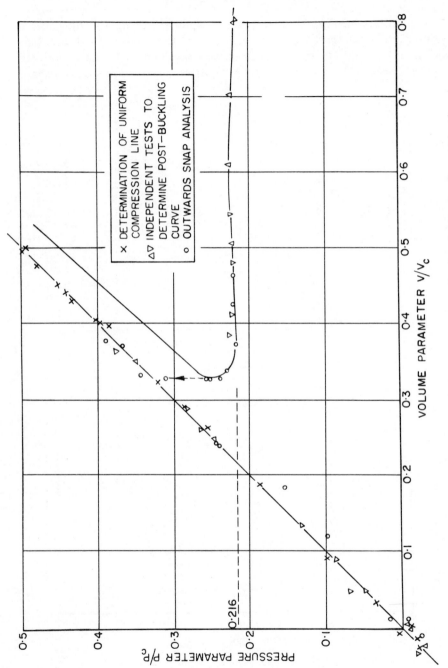

Figure 8.16 Experimental pressure–volume curves showing the results of an outwards-snap analysis[138]

inclined loading lines in the load-deflection diagram, fold catastrophes occurring where these lines touch the equilibrium path of the structure, and the dynamic jumps being forced to follow the current loading line, as shown. The enclosed area again denotes the energy lost in the hysteresis cycle, and we see that this is smaller than in the first case but larger than in the second.

The fourth arrangement shows another commonly employed form of semi-rigid loading. Here mg is the control, and since $mg = P + k\mathscr{E}$, we are again imposing inclined loading lines, as shown; analytically this arrangement is identical to the previous case.

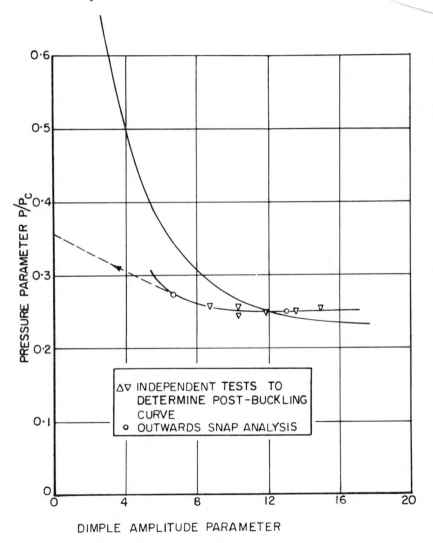

Figure 8.17 Experimental pressure–amplitude curves showing the results of an outwards-snap analysis, including comparison with a theoretical result[138]

Conceptually, these two practical semi-rigid loading devices can be viewed as 'structure plus spring' under rigid loading and 'structure plus spring' under dead loading, so that a discussion of dead and perfectly rigid loading effectively covers problems of semi-rigid loading.[18] A more general loading system could also be imagined,[137] in which a family of straight or curved loading lines is simply prescribed.

As an illustration of very rigid laboratory loading, we consider finally Thompson's experiments on a spherical shell.[138,139] The loading apparatus is shown schematically in Figure 8.13. The bore of the open manometer tube is extremely small, and thus the internal volume of the shell is essentially an imposed constant while the two taps are closed. The volume is varied incrementally by allowing a small volume of paraffin to enter or leave the system through tap A or tap B.

Results of pressure against volume showing an inwards snap analysis are shown in Figure 8.14, and the corresponding pressure–amplitude plot is shown in Figure 8.15: notice that a constant volume snap appears as an inclined line on the latter diagram. Similar outwards snap studies are represented in Figures 8.16 and 8.17. These were perhaps the first shell buckling experiments to show conclusively that the failures are 'classical' in nature, associated with folds in the equilibrium paths of imperfect specimens: previously it had often been surmised that finite dynamical disturbances were responsible for the notorious premature failures of thin elastic shells well below the eiganvalue bifurcation loads.

APPENDIX

Proof of the Lagrange and Hamilton Equations

Consider N particles with the rectangular coordinates x_α, where α ranges from 1 to $3N$.

Newton's Second Law for a typical particle along one of its axes gives

$$F_\alpha - m_\alpha \ddot{x}_\alpha = 0 \quad \text{(no summation)} \tag{1}$$

where F_α incorporates all forces on the particle.

Multiplying by a small virtual displacement δx_α and summing over all α gives us the virtual work equation

$$(F_\alpha - m_\alpha \ddot{x}_\alpha)\delta x_\alpha = 0 \tag{2}$$

We assume that compatible displacements of the system are described by a set of n generalized coordinates q_i where n is less than (or equal to) $3N$. Then we can write

$$x_\alpha = x_\alpha(q_i) \tag{3}$$

to give the first variation

$$\delta x_\alpha = (\partial x_\alpha / \partial q_i)\delta q_i \tag{4}$$

We have now the *Principle of Virtual Work* for compatible displacements

$$(F_\alpha - m_\alpha \ddot{x}_\alpha)(\partial x_\alpha / \partial q_i)\delta q_i = 0 \tag{5}$$

where there will be no contribution from smooth workless constraints.

We define the generalized forces

$$F_i = F_\alpha(\partial x_\alpha / \partial q_i) \tag{6}$$

and introducing the *Total Potential Energy* function $V(q_i)$ we can divide them into conservative and non-conservative parts,

$$F_i = -\frac{\partial V}{\partial q_i} + Q_i \tag{7}$$

195

giving our final virtual work equation

$$\left[\frac{\partial V}{\partial q_i} + m_\alpha \ddot{x}_\alpha (\partial x_\alpha / \partial q_i)\right]\delta q_i = Q_i \delta q_i \tag{8}$$

Now writing

$$x_\alpha = x_\alpha[q_i(t)] \tag{9}$$

we form the derivatives

$$\dot{x}_\alpha = (\partial x_\alpha / \partial q_i)\dot{q}_i = \dot{x}_\alpha[q_j, \dot{q}_i] \tag{10}$$

$$\partial \dot{x}_\alpha / \partial \dot{q}_i = \partial x_\alpha / \partial q_i \tag{11}$$

$$\partial \dot{x}_\alpha / \partial q_i = (\partial^2 x_\alpha / \partial q_i \partial q_j)\dot{q}_i \tag{12}$$

The *Kinetic Energy* of the system is simply

$$T = \tfrac{1}{2}m_\alpha \dot{x}_\alpha^2 \tag{13}$$

which by virtue of equation (10) is a function both of the q_i and the $\dot{q}_j$,

$$T = T(q_i, \dot{q}_j) \tag{14}$$

Forming partial derivatives of this function we have

$$\partial T / \partial \dot{q}_s = m_\alpha \dot{x}_\alpha (\partial \dot{x}_\alpha / \partial \dot{q}_s) = m_\alpha \dot{x}_\alpha (\partial x_\alpha / \partial q_s) \tag{15}$$

$$\frac{d}{dt}\frac{\partial T}{\partial \dot{q}_s} = m_\alpha \ddot{x}_\alpha (\partial x_\alpha / \partial q_s) + \underline{m_\alpha \dot{x}_\alpha (\partial^2 x_\alpha / \partial q_i \partial q_s)\dot{q}_i} \tag{16}$$

$$\frac{\partial T}{\partial q_s} = \underline{m_\alpha \dot{x}_\alpha (\partial^2 x_\alpha / \partial q_i \partial q_s)\dot{q}_i} \tag{17}$$

Subtracting equation (17) from equation (16) the underlined terms vanish and using the virtual work equation (8) we obtain

$$\frac{d}{dt}\frac{\partial T}{\partial \dot{q}_s} - \frac{\partial T}{\partial q_s} + \frac{\partial V}{\partial q_s} = Q_s \tag{18}$$

These n second-order differential equations are the *LAGRANGE EQUATIONS* of motion of a general mechanical system and introducing the *Lagrangian Function*

$$\mathscr{L} = T - V \tag{19}$$

they can be written more compactly as

$$\boxed{\frac{d}{dt}\frac{\partial \mathscr{L}}{\partial \dot{q}_i} - \frac{\partial \mathscr{L}}{\partial q_i} = Q_i} \tag{20}$$

To proceed to the Hamilton equations we define the generalized momentum

$$p_i = \frac{\partial T}{\partial \dot{q}_i} \tag{21}$$

and observing from equation (10) that the kinetic energy is quadratic in the rates $\dot{q}_i$ we write it as

$$T = \tfrac{1}{2} T_{ij}(q_k)\dot{q}_i \dot{q}_j \tag{22}$$

where the coefficients are functions of the q_k. We then have

$$p_i = T_{ij}(q_k)\dot{q}_j \tag{23}$$

which we can in principle solve for the $\dot{q}_j$ to obtain

$$\dot{q}_i = \dot{q}_i(q_k, p_l) \tag{24}$$

We now introduce the *Hamiltonian Function*

$$\mathcal{H} = T + V \tag{25}$$

which must be viewed as a function of q_i and p_j, eliminating $\dot{q}_i$ by using equation (24),

$$\mathcal{H}(q_i, p_j) = T(q_j, \dot{q}_i(q_k, p_l)] + V(q_i) \tag{26}$$

To facilitate the proof of the Hamilton equations it is convenient to write $\mathcal{H}$ more indirectly as

$$\mathcal{H}(q_i, p_j) = p_i \dot{q}_i(q_k, p_l) - T[q_j, \dot{q}_i(q_k, p_l)] + V(q_i) \tag{27}$$

which we observe is equal to $T + V$ because the first term is simply equal to $2T$.

We now form the derivatives

$$\partial\mathcal{H}/\partial p_s = \dot{q}_s + p_i(\partial\dot{q}_i/\partial p_s) - \partial T/\partial\dot{q}_i(\partial\dot{q}_i/\partial p_s) \tag{28}$$
$$\partial\mathcal{H}/\partial q_s = p_i(\partial\dot{q}_i/\partial q_s) - \partial T/\partial q_s - \partial T/\partial\dot{q}_i(\partial\dot{q}_i/\partial q_s) + \partial V/\partial q_s \tag{29}$$

and observe that a number of terms vanish by virtue of the definition of p_i in the equation (21).

So we have

$$\partial\mathcal{H}/\partial p_s = \dot{q}_s \tag{30}$$

and using the Lagrange equations (18),

$$\partial\mathcal{H}/\partial q_s = Q_s - \dot{p}_s \tag{31}$$

Setting $Q_s = 0$ for a conservative system, we now have the *CANONICAL EQUATIONS OF HAMILTON*

$$\boxed{\dot{q}_s = \frac{\partial\mathcal{H}}{\partial p_s}, \quad \dot{p}_s = -\frac{\partial\mathcal{H}}{\partial q_s}} \tag{32}$$

These are $2n$ first-order differential equations of motion.

We can finally study the time variation of $\mathcal{H}$ during a motion of the system by writing

$$d\mathcal{H}/dt = \partial\mathcal{H}/\partial q_s(dq_s/dt) + \partial\mathcal{H}/\partial p_s(dp_s/dt) = -\dot{p}_s\dot{q}_s + \dot{q}_s\dot{p}_s = 0 \tag{33}$$

We see that $\mathcal{H} = T + V$ is a constant during any motion, which proves the *conservation of energy* for our technically defined conservative system.

References

1. Thompson, J. M. T., and Hunt, G. W., *A General Theory of Elastic Stability*, Wiley, London, 1973.
2. Thompson, J. M. T., *Instabilities and Catastrophes in Science and Engineering*, Wiley, Chichester, 1982.
3. Thom, R., *Structural Stability and Morphogenesis*, translated from the French by D. H. Fowler, Benjamin, Reading, 1975.
4. Zeeman, E. C., *Catastrophe Theory: Selected Papers* 1972–1977, Addison-Wesley, London, 1977.
5. Poston, T., and Stewart, I., *Catastrophe Theory and its Applications*, Pitman, London, 1978.
6. Golubitsky, M., and Schaeffer, D., A theory for imperfect bifurcation via singularity theory, *Commun. Pure Appl. Math.* **32**, 21 (1979).
7. Thompson, J. M. T., and Hunt, G. W. (eds), *Collapse : The Buckling of Structures in Theory and Practice*, Cambridge University Press, Cambridge, 1983 (Proceedings of the IUTAM Symposium, University College London, 31 August 1982).
8. Synge, J. L., and Griffith, B. A., *Principles of Mechanics*, McGraw-Hill, New York, 1959.
9. La Salle, J., and Lefschetz, S., *Stability by Liapunov's Direct Method with Applications*, Academic Press, New York, 1961.
10. Leipholz, H., *Stability of Elastic Systems*, Sijthoff and Noordhoff, Alphen, 1980.
11. Koiter, W. T., On the instability of equilibrium in the absence of a minimum of the potential energy, *Proc. K. ned. Akad. Wet.*, Series B, **68**, 107 (1965).
12. Koiter, W. T., On the Stability of Elastic Equilibrium, Dissertation, Delft, Holland, 1945. (An English translation is now available as *NASA, Tech. Trans.*, **F10**, 833, 1967.)
13. Thompson, J. M. T., Tan, J. K. Y., and Lim, K. C., On the topological classification of post-buckling phenomena, *J. Struct. Mech.*, **6**, 383 (1978).
14. Lyttleton, R. A., *The Stability of Rotating Liquid Masses*, Cambridge University Press, Cambridge, 1953.
15. Nicolis, G., and Prigogine, I., *Self-Organization in Non-Equilibrium Systems : From Dissipative Structures to Order through Fluctuations*, Wiley, New York, 1977.
16. Katz, J., On the number of unstable modes of an equilibrium II, *Mon. Not. R. Astr. Soc.*, **189**, 817 (1979).
17. Katz, J., Stability limits for isothermal cores in globular clusters, *Mon. Not. R. Astr. Soc.*, **190**, 497 (1980).
18. Thompson, J. M. T., Stability predictions through a succession of folds, *Phil. Trans. Roy. Soc. Lond.*, A, **292**, 1–23 (1979).
19. Macmillan, N. H., and Kelly, A., The mechanical properties of perfect crystals, *Proc. R. Soc.*, A, **330**, 291 (1972).

20. Thompson, J. M. T., and Shorrock, P. A., Bifurcational instability of an atomic lattice, *J. Mech. Phys. Solids*, **23**, 21 (1975).

21. Thompson, J. M. T., and Shorrock, P. A., Hyperbolic umbilic catastrophe in crystal fracture, *Letter to Nature*, **260**, 598 (1976).

22. Hill, R., Constitutive branching in elastic materials, *Math. Proc. Camb. Phil. Soc.*, **92**, 167 (1982).

23. Michael, D. H., Meniscus stability, *Ann. Rev. Fluid Mech.*, **13**, 189 (1981).

24. Taylor, G. I., Disintegration of water drops in an electric field, *Proc. R. Soc. Lond.*, *A*, **280**, 383 (1964).

25. Berry, M. V., Cusped rainbows and incoherence effects in the rippling-mirror model for particle scattering from surfaces, *J. Phys.*, *A*, **8**, 566 (1975).

26. Berry, M. V., Waves and Thom's theorem, *Adv. Phys.*, **25**, 1, 1976.

27. Berry, M. V., Focusing and twinkling: critical exponents from catastrophes in non-gaussian random short waves, *J. Phys.*, *A*, **10**, 2061 (1977).

28. Nye, J. F., Optical caustics in the near field from liquid drops, *Proc. Roy. Soc. Lond. A*, **361**, 21 (1978).

29. Nye, J. F., Optical caustics from liquid drops under gravity: observations of the parabolic and symbolic umbilics, *Phil. Trans. Roy. Soc. A.*, **292** (1387), 25 (1979).

30. Berry, M. V., Nye, J. F., and Wright, F. J., The elliptic umbilic diffraction catastrophe, *Phil. Trans. Roy. Soc.*, **291** (1382), 453 (1979).

31. Berry, M. V., and Mackley, M. R., The six roll mill: unfolding an unstable persistently extensional flow, *Phil. Trans. R. Soc. Lond.*, *A*, **287**, 1 (1977).

32. Benjamin, T. B., Bifurcation phenomena in steady flows of a viscous fluid, I Theory, II Experiments, *Proc. R. Soc. Lond.*, *A*, **359**, 1–26 and 27–43, 1978.

33. Nye, J. F., and Thorndike, A. S., Events in evolving three-dimensional vector fields, *J. Phys.* *A*, **13**, 1 (1980).

34. Hutchinson, H. J., Nye, J. F., and Salmon, P. S., The classification of isotropic points in stress fields, *J. Struct. Mech.*,

35. Thompson, J. M. T., and Hunt, G. W., The instability of evolving systems, *Interdisciplinary Science Reviews*, **2**, 240 (1977).

36. Stewart, I., Applications of catastrophe theory to the physical sciences, *Physica (D : Nonlinear Phenomena)*, **2D**, 245 (1981).

37. Saunders, P. T., *An Introduction to Catastrophe Theory*, Cambridge University Press, Cambridge, 1980.

→ 38. Gilmore, R., *Catastrophe Theory for Scientists and Engineers*, Wiley, New York, 1981.

39. Poincaré, H., Sur l'equilibre d'une masse fluide animée d'un mouvement de rotation, *Acta. Math.*, **7**, 259 (1885).

40. Thompson, J. M. T., A general theory for the equilibrium and stability of discrete conservative systems, *Z. angew. Math. Phys.*, **20**, 797 (1969).

41. Thompson, J. M. T., Basic theorems of elastic stability. *Int. J. Engng. Sci.*, **8**, 307 (1970).

42. Chillingworth, D. R. J., A problem from singularity theory in engineering, Lecture to Symposium on Non-linear Mathematical Modelling, University of Southampton, August 1976.

43. Sewell, M. J., On the connexion between stability and the shape of the equilibrium surface, *J. Mech. Phys. Solids*, **14**, 203 (1966).

→ 44. Huseyin, K., *Non-linear Theory of Elastic Stability*, Noordhoff, Leyden, 1974.

45. Thompson, J. M. T., and Hunt, G. W., Towards a unified bifurcation theory, *J. Appl. Math. Phys. (Z.A.M.P.)*, **26**, 581 (1975).

46. Thompson, J. M. T., and Hunt, G. W., A bifurcation theory for the instabilities of optimization and design, *Synthese*, **36**, 315 (1977).

47. Hutchinson, J. W., and Koiter, W. T., Post-buckling theory, *Appl. Mech. Rev.*, **23**, 1353 (1970).

48. Budiansky, B., Theory of buckling and post-buckling behaviour of elastic structures, *Advances in Applied Mechanics*, **14**, Academic Press, New York, 1974.
49. Hutchinson, J. W., Plastic buckling. *Advances in Applied Mechanics*, **14**, Academic Press, New York, 1974.
50. Koiter, W. T., Current trends in the theory of buckling, in *Buckling of Structures* (ed. B. Budiansky), Springer-Verlag, Berlin, 1976.
51. Budiansky, B., and Hutchinson, J. W., Buckling: progress and challenge, in *Trends in Solid Mechanics* (eds J. F. Besseling and A. M. A. Van der Heijden), The University Press, Delft, 1979. (*Proceedings of the Symposium dedicated to the 65th Birthday of W. T. Koiter.*)
52. Koiter, W. T., Forty years in retrospect, the bitter and the sweet, in *Trends in Solid Mechanics* (eds J. F. Besseling and A. M. A. Van der Heijden), The University Press, Delft, 1979. (*Proceedings of the Symposium dedicated to the 65th Birthday of W. T. Koiter.*)
53. Roorda, J., Stability of structures with small imperfections, *J. Engng. Mech. Div. Am. Soc. civ. Engrs*, **91**, 87 (1965).
54. Roorda, J., The buckling behaviour of imperfect structural systems, *J. Mech. Phys. Solids*, **13**, 267 (1965).
55. Roorda, J., On the buckling of symmetric structural systems with first and second order imperfections, *Int. J. Solids Structures*, **4**, 1137 (1968).
56. Roorda, J., *Buckling of Elastic Structures*, Special Publications Series, Solid Mechanics Division, University of Waterloo Press, Waterloo, 1980.
57. Thompson, J. M. T., Catastrophe theory and its role in applied mechanics, *Proc. 14th I.U.T.A.M. Congress*, Delft, August 1976. North-Holland, Amsterdam, 1976/7.
58. Smale, S., Structurally stable systems are not dense, *Amer. J. Math.*, **88**, 491 (1966).
59. Chillingworth, D. R. J., *Differential Topology with a View to Applications*, Research Notes in Mathematics, **9**, Pitman, London, 1976.
60. Andronov, A. A., and Pontryagin, L. S., Coarse systems, *Dokl. Akad. Nauk. SSSR* **14**, 247 (1937): also in Sobraniye trudov A. A. Andronov, p. 181, *Izd. Akad. Nauk SSSR* (1956).
61. Poston, T., Various catastrophe machines, in *Structural Stability, the Theory of Catastrophes, and Applications in the Sciences*, Lecture Notes in Mathematics **525** (ed. P. J. Hilton), Springer, Berlin, 1976, pp. 111–26.
62. Thompson, D'Arcy W., *On Growth and Form*, Cambridge University Press, Cambridge, 1971.
63. Wassermann, G., Stability of unfoldings in space and time, *Acta Mathematica*, **135**, 57 (1975).
64. Wassermann, G., (r, s)-stable unfoldings and catastrophe theory, *Structural Stability, The Theory of Catastrophes, and Applications in the Sciences* (ed. P. Hilton), Lecture Notes in Mathematics, **525**, Springer, Berlin, 1976.
65. Golubitsky, M., and Schaeffer, D., An analysis of imperfect bifurcation, *Annals New York Academy of Sciences*, **316**, 127 (1979).
66. Golubitsky, M., and Schaeffer, D., Imperfect bifurcation in the presence of symmetry, *Commun. Math. Phys.*, **67**, 205 (1979).
67. Thompson, J. M. T., On the convention of a cusp in elastic stability, *J. Mech. Phys, Solids*, **31**, 205 (1983).
68. Chillingworth, D. R. J., Universal bifurcation problems in mechanics of solids, in *The Rodney Hill 60th Anniversary Volume* (eds H. G. Hopkins and M. J. Sewell), Pergamon, Oxford, 1981.
69. Marsden, J. E., Qualitative methods in bifurcation theory, *Bull. Am. Math. Soc.*, **84**, 1125 (1978).
70. Holmes, P., and Marsden, J. E., Qualitative techniques for bifurcation analysis of complex systems, *Annals New York Academy of Sciences*, **316**, 608 (1979).

71. Holmes, P. J., A nonlinear oscillator with a strange attractor, *Phil. Trans. Roy. Soc. Lond., A*, **292**, 419 (1979).

72. Thompson, J. M. T., and Ghaffari, R., Complex dynamics of bilinear systems: bifurcational instabilities leading to chaos, in *Collapse : The Buckling of Structures in Theory and Practice* (eds J. M. T. Thompson and G. W. Hunt), Cambridge University Press, Cambridge, 1983. (*Proceedings of the I.U.T.A.M. Symposium*, University College London, 31 August 1982.)

73. Niwa, Y., Watanabe, E., and Isami, H., Catastrophe analysis of structures by discretization and modal transforms, 15*th Inter. Congr. of Theoretical & Applied Mechanics*, Toronto, August, 1980.

74. Koiter, W. T., Post-buckling analysis of a simple two-bar frame, in *Recent Progress in Applied Mechanics* (eds B. Broberg, J. Hult, and F. Niordson), Almquist and Wiksell, Stockholm, 1967.

75. Chillingworth, D., The catastrophe of a buckling beam, in *Dynamical Systems, Warwick 1974* (ed. A. Manning), Lecture Notes in Mathematics, **468**, Springer-Verlag, Berlin, 1975.

76. Thompson, J. M. T., Bifurcational aspects of catastrophe theory, Proc. Conf. on Bifurcation Theory and Applications in Scientific Disciplines. New York, October 1977. *Annals, New York Acad. Sci.*, **316**, 553 (1979).

77. Hui, D., and Hansen, J. S., The swallowtail and butterfly cuspoids and their application in the initial post-buckling of single-mode structural systems, *Q. Appl. Math.*, **38**, 17 (1980).

78. Thompson, J. M. T., Tulk, J. D., and Walker, A. C., An experimental study of imperfection-sensitivity in the interactive buckling of stiffened plates, in *Buckling of Structures* (ed. B. Budiansky), Springer-Verlag, Berlin, 1976.

79. Sewell, M. J., The static perturbation technique in buckling problems, *J. Mech. Phys. Solids*, **13**, 247 (1965).

80. Thompson, J. M. T., Discrete branching points in the general theory of elastic stability, *J. Mech. Phys. Solids*, **13**, 295 (1965).

81. Hunt, G. W., Symmetries of elastic buckling, *Eng. Struct.*, **4**, 21 (1982).

82. Thompson, J. M. T., and Hunt, G. W., On the buckling and imperfection-sensitivity of arches with and without pre-stress, Int. J. Solids Structures, **19**, 445 (1983).

83. Biezeno, C. B., and Grammel, R., *Engineering Dynamics*, Vol. 2, Blackie, London, 1960.

84. Harrison, H. B., Post-buckling behaviour of elastic circular arches, *Proc. Inst. civ. Engrs*, Part 2, **65**, 283, (1978).

85. Croll, J. G. A., and Walker, A. C., *Elements of Structural Stability*, Macmillan, London, 1972.

86. Thompson, J. M. T., and Gaspar, Zs., A buckling model for the set of umbilic catastrophes, *Math. Proc. Camb. Phil. Soc.*, **82**, 497 (1977).

87. Hunt, G. W., Reay, N. A., and Yoshimura, T., Local diffeomorphisms in the bifurcational manifestations of the umbilic catastrophes, *Proc. R. Soc. Lond., A*, **369**, 47 (1979).

88. Hansen, J. S., Some two-mode buckling problems and their relation to catastrophe theory, *AIAA Journal*, **15**, 1638 (1977).

89. Chilver, A. H., Coupled modes of elastic buckling, *J. Mech. Phys. Solids*, **15**, 15 (1967).

90. Johns, K. C., Imperfection sensitivity of coincident buckling systems, *Int. J. Nonlin. Mech.*, **9**, 1 (1974).

91. Supple, W. J. (ed.), *Structural Instability*, IPC Science and Technology Press, Guildford, 1973.

92. Hunt, G. W., Imperfections and near-coincidence for semi-symmetric bifurcations, Proc. Conf. on Bifurcation Theory and Applications in Scientific Disciplines, New York, October 1977. *Annals, New York Acad. Sci.*, **316**, 572 (1979).

93. Hunt, G. W., Imperfection-sensitivity of semi-symmetric branching, *Proc. R. Soc. Lond., A*, **357**, 193 (1977).
94. Supple, W. J., Initial post-buckling behaviour of a class of elastic structural systems, *Int. J. Nonlin. Mech.*, **4**, 23 (1969).
95. Ho, D., Buckling load of non-linear systems with multiple eigenvalues, *Int. J. Solids Structures*, **10**, 1315 (1974).
96. Samuels, P., Bifurcation and limit point instability of dual eigenvalue third order systems, *Int. J. Solids Structures*, **16**, 743 (1980).
97. Samuels, P., The relationship between postbuckling behaviour at coincident branching points and the geometry of an umbilic point of the energy surface, *J. Struct. Mech.*, **7**, 297 (1979).
98. Tvergaard, V., Imperfection-sensitivity of a wide integrally stiffened panel under compression, *Int. J. Solids Structures*, **9**, 177 (1973).
99. Gaspar, Zs., Imperfection-sensitivity and catastrophe theory, in Proc. IUTAM Symp., *Collapse: the Buckling of Structures in Theory and Practice* (eds J. M. T. Thompson and G. W. Hunt), Cambridge University Press, Cambridge, 1983.
100. Born, M., and Huang, K., *Dynamical Theory of Crystal Lattices*, Oxford University Press, Oxford, 1954.
101. Hunt, G. W., and Williams, K. A. J., Closed-form and asymptotic solutions for an interactive buckling model J. Mech. Phys. Solids (to appear).
102. Hunt, G. W., and Williams, K. A. J., On truncation of the structural potential function (to be published).
103. Hunt, G. W., An algorithm for the nonlinear analysis of compound bifurcation, *Phil. Trans. R. Soc. Lond., A*, **300** (1455), 443 (1981).
104. Hui, D., and Hansen, J. S., The parabolic umbilic catastrophe and its application in the theory of elastic stability, *Q. Appl. Math.*, **39** (2), 201 (1981).
105. Arnold, V. I., Critical points of smooth functions and their normal forms, *Uspehi Mat. Nauk*, **30**, 3 (1975); *Russian Math. Surveys*, **30**, 1 (1975).
106. Callahan, J. J., The geometry of $E_6 = x^3 + y^4$, anorexia nervosa, and the method of tableaus for visualizing five dimensional objects, in *Graphic Techniques in Geometry and Topology* (ed. G. K. Francis), Proc. of Special Session, Amer. Math. Soc., Evanson, Illinois, 1977.
107. Magnus, R., and Poston, T., On the full unfolding of the von Kármán equations at a double eigenvalue, *Report of the Battelle Advanced Studies Centre*, Geneva, August, 1977.
108. Augusti, G., Stabilita di strutture elastiche elementari in presenza di grandi spostamenti, *Atti Accad. Sci. fis. mat., Napoli, Serie 3ª*, **4**, No. 5 (1964)
109. Thompson, J. M. T., and Supple, W. J., Erosion of optimum designs by compound branching phenomena, *J. Mech. Phys. Solids*, **21**, 135 (1973).
110. Sewell, M. J., A general theory of equilibrium paths through critical points, *Proc. R. Soc. Lond., A*, **306**, 201 (1968).
111. Sewell, M. J., A method of post-buckling analysis, *J. Mech. Phys. Solids*, **17**, 219 (1969).
112. Sewell, M. J., On the branching of equilibrium paths, *Proc. R. Soc. Lond., A*, **315**, 499 (1970).
113. Poston, T., and Stewart, I., *Taylor Expansions and Catastrophes*, Research Notes in Mathematics, 7, Pitman, London, 1976.
114. Schaeffer, D., and Golubitsky, M., Boundary conditions and mode jumping in the buckling of a rectangular plate, *Commun. Math. Phys.*, **69**, 209 (1979).
115. Golubitsky, M., Marsden, J., and Schaeffer, D., Bifurcation problems with hidden symmetries, *Report PAM-117*, Center for Pure and Applied Mathematics, University of California, Berkeley, January 1983 (to be published).
116. Timoshenko, S. P., and Gere, J. M., *Theory of Elastic Stability*, McGraw-Hill, New York, 1961.

117. Koiter, W. T., and Pignataro, M., A general theory for the interaction between local and overall buckling of stiffened panels. *Rep. WTHD* 83, Delft University of Technology, Delft, 1976.

118. Calladine, C. R., The static-geometric analogy in the equations of thin shell structures, *Math. Proc. Camb. Phil. Soc.*, **82**, 335 (1977).

119. van der Neut, A., The interaction of local buckling and column failure of thin-walled compression members, *Proc. XII Internat. Cong. Appl. Mech.*, Stanford, 1968 (Springer-Verlag).

120. Thompson, J. M. T., and Lewis, G. M., On the optimum design of thin-walled compression members, *J. Mech. Phys. Solids*, **20**, 101 (1972).

121. Gilbert, R. B., and Calladine, C. R., Interaction between the effects of local and overall imperfections on the buckling of elastic columns, *J. Mech. Phys. Solids*, **22**, 519 (1974).

122. Maquoi, R., and Massonnet, Ch., Interaction between local plate buckling and overall buckling in thin-walled compression members—theories and experiments, in *Buckling of Structures* (ed. B. Budiansky), Springer-Verlag, Berlin, 1976.

123. Walker, A. C., Interactive buckling of structural components, *Sci. Prog. Oxf.*, **62**, 579 (1975).

124. Rhodes, J., and Walker, A. C. (eds), *Thin-Walled Structures*, Granada, London, 1980.

125. Rondal, J., and Maquoi, R., On the optimum design of square hollow compression members, Proc. IUTAM Symp., *Collapse : the Buckling of Structures in Theory and Practice* (eds J. M. T. Thompson and G. W. Hunt), Cambridge University Press, Cambridge, 1983.

126. Koiter, W. T., General theory of mode interaction in stiffened plate and shell structures, *Rep. WTHD* 91, Delft University of Technology, Delft, 1976.

127. Koiter, W. T., The effect of axisymmetric imperfections on the buckling of cylindrical shells under axial compression, *Proc. K. ned. Akad. Wet.*, B, **66**, 265 (1963).

128. Koiter, W. T., The nonlinear buckling problem of a complete spherical shell under uniform external pressure, *Proc. K. ned. Akad. Wet.*, B, **72**, 40 (1969).

129. Hansen, J. S., Influence of general imperfections in axially loaded cylindrical shells, *Int. J. Solids Structures*, **11**, 1223 (1975).

130. Calladine, C. R., and Robinson, J. M., A simplified approach to the buckling of thin shells, *IUTAM Conf.*, Tblisi, 1978.

131. Donnell, L. H., A new theory for the buckling of thin cylinders under axial compression and bending, *Trans. ASME*, **56**, 795 (1934).

132. Croll, J. G. A., Lower bound elasto-plastic buckling of cylinders, *Proc. Inst. civ. Engrs.*, Part 2, **71**, 235 (1981).

133. Yoshimura, Y., On the mechanism of buckling of a circular cylindrical shell under axial compression, *Reports of the Institute of Science and Technology of the University of Tokyo*, Vol. 5, No. 5, 1951 (English translation: *Technical Memorandum No. 1390 of the National Advisory Committee for Aeronautics*, Washington, DC, 1955).

134. von Karman, T., and Tsien, H. S., The buckling of spherical shells by external pressure, *J. aeronaut. Sci.*, **7**, 43 (1939).

135. Thompson, J. M. T., The rotationally-symmetric branching behaviour of a complete spherical shell, *Proc. K. ned. Akad. Wet.*, B, **67**, 295 (1964).

136. Hutchinson, J. W., Imperfection sensitivity of externally pressurized spherical shells, *J. appl. Mech.*, **34**, 49 (1967).

137. Thompson, J. M. T., Stability of elastic structures and their loading devices, *J. mech. Engng Sci.*, **3**, 153 (1961).

138. Thompson, J. M. T., The elastic instability of a complete spherical shell, *Aero. Quart.*, **13**, 189 (1962).

139. Thompson, J. M. T., The post-buckling of a spherical shell by computer analysis, in *World Conference on Shell Structures* (eds S. J. Medwadowski et al.), National Academy of Sciences, Washington, 1964.

INDEX

active coordinates, 43, 150
activity surface, 151, 165
analyticity, 151
Andronov, A. A., 54
anticlinal bifurcation
    generalized definition, 134
    imperfection-sensitivity, 131
    paths, 121
    routes, 137
    splitting parameter, 124
arch buckling
    experiments, 86
    model for, 53
    post-buckling, 91
    symmetries of, 167
    tied model for, 60
Arnold, V. I., 142, 143
asymmetric bifurcation, 46, 55, 59, 64, 86
    compound, 133
    imperfection-sensitivity, 49
asymmetric frame, 70
asymmetric model, 67
asymptotic analysis, 79
asymptotic stability, 7, 10
atomic lattice, 137
Augusti model, 144
axioms
    equilibrium, 6
    stability, 11

barrelling of cylinder, 180
basic theorems, 102, 149
beak-to-beak singularity, 85
beam formulation, 27
Benjamin, T. B., 41
Berry, M. V., 41, 58
Biezeno and Grammel, 108
bifurcational
    classification, x, 47, 57, 58

formalism, 47, 55, 60, 64, 75, 148
    parameter, 149
bilinear failure locus, 139
boundary conditions, geometric, natural,
    30
buckling
    modes for hinged strut, 19, 26
    plates and shells, 164
butterfly singularity, 88, 111, 179

Callahan, J. J., 142
canonical equations of Hamilton, 197
cantilever, vibration of, 33
catastrophe theory, ix, 42, 43, 47, 56
caustics, 41
chaotic motions, 59
characteristic determinant, 25
    hinged strut, 26
    cantilever, 34
chequerboard pattern of cylinder, 180
Chillingworth, D. R. J., 47, 54, 59, 77
Chilver, A. H., 123
circular frequency, 24
    beam or column, 32
    string, 33
classical stability definition, 7
coincident buckling modes, 42
compact geometry of umbilics, 143
complementary path, 51
compound bifurcation
    classification of asymmetric points, 134
    critical point, 15, 119
    cylinder buckling, 180
    semi-symmetric analysis, 160
    sphere buckling, 184
comprehensive bifurcation analysis, 147
conservation of energy, 4
    pendulum, 8
    proof of, 197

constraint condition, for arch, 95
contamination by passive modes, 23, 153, 166
control parameters, 40
control space, 58
convolutions in arch response, 108
coriolos forces, 40
coupled and uncoupled paths, 119
critical loads
    columns and struts, 32
    hinged strut, 19
critical state, 15
    arch, 99
    identity, 156
    stability of, 15
Croll, J. G. A., 108, 184
cross-terms, elimination of, 13
crystal fracture, 41, 139
cusp, 46, 73
    arch, 102
    cuspoid catastrophes, 57, 86
    double, 57, 89, 142, 143, 175, 180
    dual, 46, 58, 73, 81
    line of, 108
cut-off point, 59, 75, 84, 85, 111
cylindrical shell, 56, 180

damping, positive definitive, 10
dead and rigid loading, 188
de-coupled equations of motion, 24
degrees of freedom, 1
degrees of instability, 63
delay convention, 45
determinacy, 55, 151, 159, 178, 184
developmental biology, 86
diagonalization, 13
discretization, 2
dissipation of energy, 10
distinct critical points, 15, 42, 60
divergence within a cusp, 73
domes, buckling of, 86
Donnell's cylinder buckling theory, 183
double cusp, 57, 89, 142, 143, 175, 180
dual cusp, 46, 58, 73, 81
dummy-suffix convention, 3
dynamic
    disturbances on shells, 194
    fast action, 64, 73
    instabilities, ix
    snap, 63
dynamical systems theory, ix

eigenvalue analysis, 93, 98

eigenvector, local, 156
Einstein summation convention, 3
elastic foundation, 29
elasticity of loading, 190
elimination of passive controls, 53
elimination of passive coordinates, 23, 42, 43, 150
elliptic umbilic
    guyed cantilever, 116
    routes, 135
    surface, 131
end-shortening of
    column, 29
    column element as matrix, 36
    hinged strut, 18
energy surfaces, 9, 16, 19
energy theorems, 9
equations of motion, 5
equilibrium
    condition, 5
    identity, 154
Euler, L., 169
Euler
    critical load. 76, 169
    mode for stiffened plate, 175
    strut, 23, 76
exchange of stability, 46, 66
experiments on
    arch, 105
    spherical shell, 191

family of systems, 40
finite element method, 2, 27, 34
first-order work, 149, 156
fluid loading, 12
fluid mechanics, 41, 66
flutter, ix
fold, 45, 55, 60, 189
Fourier expansion, 31, 93
frame model, 53
free motions, 7
fundamental path, 45, 148

Gasper, Z., 112, 128
Gaussian curvature, 66, 173
generalized
    coordinates, 1, 7
    forces, 195
    imperfections, 147, 157
    loads, 147, 158
    momentum, 196
genericity, 54
Gilmore, R., 42

Golubitsky, M., x, 48, 57, 58, 66, 75, 86, 107, 127, 169
gravitational potential, 12
guyed cantilever, 112
  anticlinal bifurcation, 131
  general theory, 119, 130
  monoclinal bifurcation, 133
  paraclinal bifurcation, 142
  symbolic umbilic, 142
gyroscopic systems, 1, 40

Hamilton equations, 7, 195
Hamiltonian function, 197
Hansen, J. S., 89, 112, 142
harmonic analysis, of column, 31
Harrison, H. B., 108
Heath, N., 107
Hill, R., 41
hill-top bifurcation
  arch, 104
  crystal fracture, 137, 139
  route, 137
history-dependence, 73
Ho, D., 126
Holmes, P.J., 59
holonomic system, 2
homeoclinal bifurcation, 118
  equilibrium paths, 122
  generalized definition, 135
  imperfection-sensitivity, 128, 129
  route, 137
  splitting parameter, 125
Hooke's law, 91
Hui, D., 89, 142
Huseyin, K., 47, 158
Hutchinson, J. W., 180, 185
hyperbolic umbilic
  guyed cantilever, 116
  route, 136
  spherical shell, 184
  surface, 127, 128, 129, 132
hysteresis
  cusp, 73
  dead and rigid loading, 189
  loop, 63
  point (non-degenerate), 59, 85

ignorable controls, 56
imperfections, 40
imperfection ray, 157
imperfection-sensitivity, 48
  arch, 104
  coincident eigenvalues, 126

cylindrical shell, 182
  semi-symmetric bifurcations, 118
  stiffened plate, 177
  surfaces, 126
implicit function theorem, 43, 153
inertial mass, 11
inextensible arch, 91
infinitesimal disturbances, 7
inflexion point, 44, 63
interactive buckling, 112, 164, 175, 184
intrinsic fourth derivative, 23
intrinsic perturbation scheme, 95, 150
invariant statements, 13, 14

Johns, K. C., 125
Jordan, K., 111

Karman, T. von, 185
Karman plate equations, 144, 171
kinetic energy, 3, 11, 29, 196
Koiter, W. T., 10, 42, 48, 70, 133, 170, 171, 173, 175, 177, 179, 180, 182, 184, 186, 188

laboratory loading, 188, 190
Lagrange energy theorem, 1, 9, 10
Lagrange equations, ix, 1, 3, 4, 24
  proof, 195
  strut, 31
Lagrange multiplier, classification, 133
Lagrangian function, 4, 24, 196
La Salle and Lefschetz, 7
Legendre functions, 185, 186
Lennard-Jones interatomic potential, 137
Liapunov stability definition, 1, 7
limit point, 45, 49, 63
linear
  buckling of struts, 27
  eigenvalue analysis, 21, 24
  transformation, 13
  vibrations, 1, 23, 27
lip singularity, 85
loads, 40
local mode of stiffened plate, 175
lower bound, 171, 184
lower buckling load, of sphere, 185

Mackley, M. R., 41
Magnus, R., 144
major and minor imperfections, 108
Marsden, J. E., 59
Maxwell convention, 45
membrane stresses in plate, 171

meniscus studies, 41
metastability, 9, 45
m-fold bifurcation, 150
minimum, condition for, 12
modal analysis, 2
modal expansions, 29, 76
mode interactions, 140
modelling, mathematical, 54
monoclinal bifurcation, 118
   equilibrium paths, 120
   generalized definition, 135
   imperfection-sensitivity, 127, 132
   route, 137
   splitting parameter, 123
morphogenesis, 58

natural path, 51
near-coincidence of critical loads, 122
necessary and sufficient conditions
   equilibrium, 5
   stability, 11
Newton's Laws, 4, 195
non-conservative systems, ix, 12, 41, 195
normal modes, 1, 24, 26
null matrix, 153
numerical search procedure, 162
Nye, J. F., 41

oblique axes, 13
observations, experimental, 55
off-set of cusp, 106
off-set of load, 104
once-and-for-all disturbance, 7
optics, 41
optimization, 145, 164, 167
orthogonality, 31, 76, 95

panel, 177
parabolic umbilic, 140, 159, 164
   cylindrical shell, 183
   guyed cantilever, 118
   stiffened plate, 178
paraclinal bifurcation, 168
   equilibrium paths, 141
   imperfection-sensitivity, 142
passive coordinates, elimination of, 23, 42, 43, 150, 165
pendulum analysis, 4, 6, 8, 12
perfect and imperfect systems, 48
perturbation analysis, 23, 96, 150, 153
perturbation parameter, 154
perturbed bifurcation, 47
phase space, 7, 8

photo-elasticity, 41
Pignataro, M., 170, 177
pitch-fork bifurcation, 46, 75, 76
planets, 40
   double cusp, 144
plate buckling, 164
   double cusp, 144
   model for, 77
   post-buckling analysis, 170
Poincaré, H., 46, 66
Pontryagin, L. S., 54
positive definite form, 4, 14
positive semi-definite form, 15
Poston, T., 41, 55, 56, 58, 85, 89, 111, 116, 144, 151, 184
post-buckling stiffness, 79
pre-stress, 91
primary path, 45, 148
principal coordinates, 13, 19, 165
principal imperfections, 125, 126, 160

quadratic forms, 12

random disturbances, 7
rank, co-rank, 43, 150
Rayleigh–Ritz analysis, 30, 42, 94, 166, 169
reduced Euler load, 89
rigid loading, 63, 64, 81, 108, 185, 188
rigid platens, 171
ring buckling, 170
Roorda, J., 53, 70, 86, 104, 105, 108
root structure of cubic, 133
rotating liquid masses, 40
routes through
   catastrophes, 58
   cusp, 83
   fold, 72
   umbilics, 135

saddle point, 8, 16
Samuels, P., 126
Saunders, P. T., 42
Schaeffer, D. G., x, 48, 57, 58, 66, 75, 86, 107, 127
scleronomic system, 3
secondary bifurcation, 123, 124
   Augusti model, 145
   general analysis, 156
   identity, 157
secondary paths, 149

segregation
  of controls, 155
  of coordinates, 150
semi-rigid loading, 188
semi-symmetric bifurcation, 118
seven elementary catastrophes, 56
Sewell, M. J., 47, 95, 150
shallow shell theory, 180
Shanley column, 90
shear lag, 178
shell buckling, 164
  classical nature of failure, 194
  model for, 81
  scatter of buckling loads, 48
ship stability, 88
Shorrock, P. A., 137
singularity theory, ix
  without potential, 73
simultaneous buckling, 42, 113, 177
simultaneous diagonalization, 18, 24
skew-symmetric deformation, 20
sliding coordinates, 64, 67, 76, 81
  general analysis, 149
  plate model, 78
Smale, S., 54
snap buckling, 63, 103, 109
spherical shell, 56, 184
  experiments on, 191
  incomplete unfolding of, 137, 138
splitting lemma, 153
splitting parameters, 89, 114, 118, 122,
    140, 158, 160
stability
  coefficient, 13, 19, 99
  critical state, 21
  definition, 7
  determinant, 15, 21
  with respect to a coordinate, 19
stable-symmetric bifurcation, 23, 46, 75
  imperfections, 51
  plate, 173
stars, 40
state space, 1
Stewart, I., 41, 42, 55, 56, 58, 85, 89, 111
stiffened plates, 90, 128, 175
stiffened structures, 175
stiffness, in post-buckling, 173
strain energy
  beam, 28
  hinged strut, 17
  matrix for beam element, 36
  prestress, 91
strange attractors, 59

stress function, 41
structural stability, topographical sense, ix,
    48, 53
strut, 27, 88, 169
sub-critical bifurcation, 46
succession of folds, 64
sum of squares, 13
super-critical bifurcation, 46
Supple, W. J., 126, 144, 145
swallow-tail singularity, 86
symbolic umbilic, 118, 142
symmetry, 164, 167
  breaking, 51, 168
  criteria, 168
  cylinder buckling, 181, 184
  double, 175
  hidden, 169
  interactive buckling, 112
  plates, 174, 175
  section of, 125
Synge and Griffith, 2

Taylor expansion, 6, 12
tensor notation, 3
theorems, basic, 46, 102, 149
thermodynamics, 40
Thom, R., x, 47, 54, 56, 58, 73, 85, 127,
    135, 142
Thompson, D'Arcy W., 56
thoroughly stable (unstable), 14
tied arch, 137, 139
tilt of arch, 107
tilting of cusp, 59, 75, 85, 107
total energy contours, for pendulum, 8
total potential energy, 3
transcritical bifurcation, 46
transformed energy function, 149
Tsien, H. S., 185
Tvergaard, V., 128, 129
two-hinged strut, 17, 25
two-thirds power law, 51, 73, 81, 104

umbilic bracelet, 113, 115, 116
umbilic catastrophes, 57, 112
unfolding, 55
  of cylinder, 184
universal bifurcation diagrams, 51, 52
unstable–symmetric bifurcation, 23, 44, 51,
    75, 81

variational mechanics, 11
vector field, 7
Virtual Work Principle, 195

viscous damping, 1

Walker, A. C., 108
Wassermann, G., 57
Williams, K. A. J., 183

Yoshimura pattern for cylinder, 183, 184

Zeeman, E. C., x, 56, 58, 73, 77, 86, 88, 89,
    91, 95, 113, 116, 143, 144.